Rheinwerk Computing

The Rheinwerk Computing series offers new and established professionals comprehensive guidance to enrich their skillsets and enhance their career prospects. Our publications are written by the leading experts in their fields. Each book is detailed and hands-on to help readers develop essential, practical skills that they can apply to their daily work.

Explore more of the Rheinwerk Computing library!

Michael Kofler

Scripting: Automation with Bash, PowerShell, and Python

2024, 470 pages, paperback and e-book
www.rheinwerk-computing.com/5851

Philip Ackermann

Full Stack Web Development

2023, 740 pages, paperback and e-book
www.rheinwerk-computing.com/5704

Philip Ackermann

JavaScript: The Comprehensive Guide

2022, 1292 pages, paperback and e-book
www.rheinwerk-computing.com/5554

Johannes Ernesti, Peter Kaiser

Python 3: The Comprehensive Guide

2022, 1036 pages, paperback and e-book
www.rheinwerk-computing.com/5566

Veit Steinkamp

Python for Engineering and Scientific Computing

2024, 511 pages, paperback and e-book
www.rheinwerk-computing.com/5852

www.rheinwerk-computing.com

Thomas Theis

Getting Started with
Python

Editor Megan Fuerst
Acquisitions Editor Hareem Shafi
German Edition Editor Anne Scheibe
Translation Winema Language Services, Inc.
Copyeditor Yvette Chin
Cover Design and Illustration Mai Loan Nguyen Duy
Layout Design Vera Brauner
Production Graham Geary
Typesetting III-Satz, Germany
Printed and bound in Canada, on paper from sustainable sources

ISBN 978-1-4932-2586-6
© 2024 by Rheinwerk Publishing, Inc., Boston (MA)
1st edition 2024
8th German edition published 2024 by Rheinwerk Verlag, Bonn, Germany

Library of Congress Cataloging-in-Publication Data:
Names: Theis, Thomas, author.
Title: Getting started with Python / by Thomas Theis.
Description: 1st edition. | Bonn ; Boston : Rheinwerk Publishing, 2024. |
  Includes index.
Identifiers: LCCN 2024018654 | ISBN 9781493225866 (hardcover) | ISBN
  9781493225873 (ebook)
Subjects: LCSH: Python (Computer program language)
Classification: LCC QA76.73.P98 T47 2024 | DDC 005.13/3--dc23/eng/20240509
LC record available at https://lccn.loc.gov/2024018654

Contents at a Glance

Contents

5   Advanced Programming

6   Object-Oriented Programming

7   Various Modules

8   Files

247

9 Databases

10   User Interfaces

11   User Interfaces with PyQt   361

Appendices

Materials for This Book

The following resources are available for you to download from this book's webpage:

- **All sample programs**
- **Solutions to the exercises**

Go to *https://www.rheinwerk-computing.com/5876* and scroll down to the **Product supplements**. You will see downloadable files. Click the **Download** button to start the download process. Depending on the size of the file (and your internet connection), you may need some time for the download to complete.

Chapter 1
Introduction

In this chapter, I will briefly introduce you to Python. You'll learn about the advantages of Python and how to install Python on Windows, Ubuntu Linux, and macOS.

1.1   Advantages of Python

Python is an extremely easy programming language to learn and is ideal for getting started in the world of programming. Despite its simplicity, this language also offers the possibility of writing complex programs for a wide range of applications.

Python is particularly suitable for the rapid development of extensive applications and combines the following advantages for this purpose:

- **A simple, unambiguous syntax**
  Python is an ideal programming language for anyone starting out in programming. It is limited to simple, clear statements and often to a single possible solution. This language can be quickly learned and memorized and becomes familiar during the development of programs.

- **Clear structures**
  Python requires you to write in an easily readable structure. The arrangement of program lines also determines the logical structure of the program.

- **Reuse of code**
  Modularization, that is, breaking down a problem into sub-problems and then combining the partial solutions into an overall solution, is quite easy in Python. The existing partial solutions can easily be used for further tasks, so you'll soon have an extensive pool of modules at your disposal.

- **Improved troubleshooting**
  Python versions 3.10 to 3.12 include many improvements to error messages. Error messages are now more informative and accurate and therefore help better with troubleshooting. Suitable suggestions for rectifying an error are often already made.

- **Object-oriented processing**
  In Python, all data is stored as objects. This approach leads to a uniform treatment of objects of different types. The physical storage of Python objects is automatic. During development, you don't need to worry about reserving and releasing suitable memory areas.

- **Interpreter/compiler**
  Python programs are interpreted directly. They don't have to be compiled and bound first, which enables frequent, rapid switching between the coding and test phases.

- **Independence from the operating system**
  Both programs that are operated from the command line, and programs with graphical user interfaces (GUIs) can be used on different operating systems (Windows, Linux, macOS) without new development and adaptation.

- **Expansion possibilities**
  Numerous libraries are either already integrated or can be integrated very easily. The libraries contain useful, specialized functions from many areas.

1.2   Prevalence of Python

Due to its many advantages, Python has been one of the most popular programming languages for many years. Currently (as of spring 2024), Python occupies first place in various programming language rankings. Python is used in numerous large companies; here are some examples:

- The Spotify music app was created using Python.
- At Netflix, Python is increasingly being used for development, despite a choice among different systems.
- Instagram uses Python in an important part of its application.
- Google uses Python as often as possible.
- Dropbox uses Python exclusively.

1.3   Structure of This Book

This book introduces you to Python 3.12, which was released in October 2023. In this book, I place particular emphasis on having you work with Python yourself. I therefore recommend that you follow the logical thread of chapters and examples right from the start.

The first correlations are explained in Chapter 2 using simple calculations. You'll also learn how to enter a program, save it, and run it in various environments.

You'll get to know the language in a playful way. For this reason, you'll program a game that accompanies you through this book. Our game involves solving one or more mental arithmetic tasks. I introduce this game in Chapter 3, and we'll continuously expand and improve it throughout the rest of this book.

After the presentation of the different data types with their respective properties and advantages in Chapter 4, your programming knowledge is deepened in Chapter 5. Chapter 6 is dedicated to object-oriented programming (OOP) using Python, and some useful modules to supplement the programs are presented in Chapter 7.

In Chapter 8 and Chapter 9, you'll learn how to save data permanently in files or databases.

Windows as well as Ubuntu Linux and macOS offer convenient GUIs. Chapter 10 deals with GUI generation using the `tkinter` module, which represents an interface between the graphical *Tk* toolkit and Python. Chapter 11 deals with GUI generation using the `PyQt6` module, which contains the elements of *PyQt 6*. PyQt serves as an interface between the *Qt* library and Python.

I would like to thank the entire teams at Rheinwerk Verlag and Rheinwerk Publishing, notably Anne Scheibe for the original German edition and Hareem Shafi, Megan Fuerst, and Winema Language Services for their help in creating this English edition.

1.4   Exercises

The book contains numerous exercises. I recommend that you always solve them immediately as you work through the book. In this way, you can test your knowledge before moving on to the next topic. You'll find the solutions to the exercises together with the sample programs in the materials accompanying the book. In doing so, note the following:

- Many correct solutions may exist for any given problem. Your solution may not look exactly like the one provided in this book, which is not a problem. You should rather consider the provided solution as a suggestion and an alternative to the one you came up with.

- When solving tasks yourself, you'll certainly make one or two mistakes; don't let this discourage you, because you can only learn from mistakes. The approaches suggested in this book are how you'll really learn Python—not just by reading programming rules.

1.5   Installation on Windows

Python is a freely available programming language that can be used on various operating systems. You can download the latest Python versions from the Python website at *https://www.python.org*. As I write this book (in spring 2024), the *python-3.12.2-amd64.exe* file is the most recent. You'll need to call this file for installation. First, you must confirm that you want to install a program that does not originate from the Microsoft Store.

Select the **Add Python to Path** option so that you can later start Python programs at the command line level from any directory.

Select the **Customize Installation** option. Otherwise, leave all the default options set. This approach applies in particular to the *pip* package management program, which you can use to install additional modules and programs later (see also Appendix A, Section A.1).

Then, select the installation directory *C:\Python* in the **Customize install location** field.

You'll then see an entry for **Python 3.12** in the **Start** menu, as shown in Figure 1.1. If you want to change certain installation settings afterwards, run the installation file again.

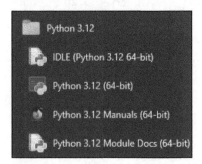

Figure 1.1 Start Menu with Entry for Python 3.12

Now, in your **Start** menu, the Integrated Development and Learning Environment (IDLE) program is a development environment that itself has been written in Python. Since you'll use IDLE to write your programs, I recommend dragging a shortcut to IDLE onto the desktop.

1.6   Installation on Ubuntu Linux

Ubuntu Linux 23.10 is used in this book as representative of other Linux distributions. Python 3 is already installed on Ubuntu Linux and must not be uninstalled. In a terminal, use the `python3 -V` command (with an uppercase "V") to determine your current version number.

To install the IDLE development environment, enter the following command in a terminal: `sudo apt install idle3`. You can then start the IDLE program via the `idle` command.

At the time of this writing (spring 2024), Python version 3.11.6 is still used on Ubuntu Linux. For this reason, the new features of Python version 3.12 mentioned in this book cannot be used yet, but this will change in the near future.

1.7   Installation on macOS

You can download the latest versions of Python from the official website at *https://www.python.org*. Currently (as of spring 2024), the latest version is the *python-3.12.2-macos11.pkg* file for macOS.

Double-click on this file to start the installation process. If you don't make any changes, Python will be installed in the *Programs/Python 3.12* directory, where you'll find an entry for the IDLE development environment. Drag this entry onto the desktop to create a shortcut.

Chapter 2
Getting Started

In this chapter, you'll use Python for the first time—initially as a calculator. You'll also learn how to enter, save, and run a program.

2.1 Python as a Calculator

You can initially use Python as a simple pocket calculator. This straightforward example makes it easier to get started with Python.

2.1.1 Entering Calculations

Call the Integrated Development and Learning Environment (IDLE) for Python as described at the end of the various installations in Chapter 1. The IDLE development environment can be used both as an editor for entering programs and as a simple calculator.

An example IDLE screen on Windows is shown in Figure 2.1.

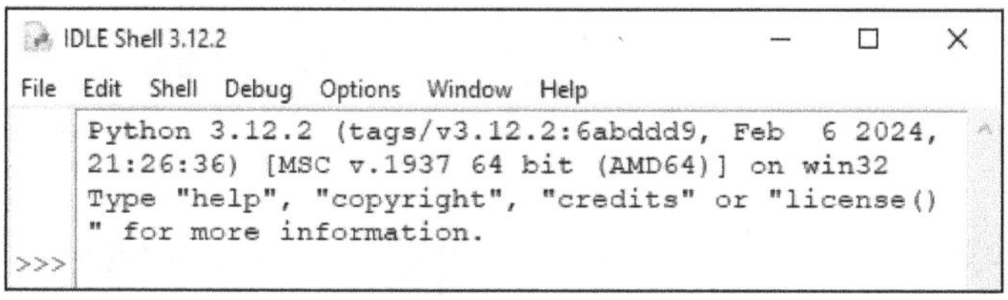

Figure 2.1 Python Development Environment IDLE

You don't need to write and start complete programs to perform small calculations. Instead, you can enter the required arithmetic operations directly at the input prompt, which is recognizable by the >>> string. Finally, pressing Enter immediately calculates and displays the result.

The calculation rules applied here in IDLE are also used in the creation of Python programs. The *input, processing, and output (IPO)* model is already applied, as in many simple programs. When using this model, you enter the data, which is processed using a calculation, and then the result is output.

> **Note**
>
> Most of the illustrations in this book were created for Python on Windows, but they apply in a similar way to Python on Ubuntu Linux and on macOS.

2.1.2 Addition, Subtraction, and Multiplication

Let's consider some simple calculations separated into IPO elements:

```
>>> 41 + 7.5
48.5
>>> 12 - 18
-6
>>> 7 * 3
21
```

The + (addition), - (subtraction), and * (multiplication) operators are used one after the other. When you press ⏎Enter, the result will appear in the line below.

You can use both integers and numbers that have decimal places for your calculations. The decimal places must be separated using a decimal point.

2.1.3 Division, Integer Division, and Modulo

The / operator is used for mathematical division, as in the following examples:

```
>>> 22 / 8
2.75
>>> 22 / -8
-2.75
```

The // operator is used to perform an integer division, as in the following examples:

```
>>> 22 // 8
2
>>> 22 // -8
-3
>>> 7 // 2.5
2.0
```

When you enter an integer division using the // operator, the result of the mathematical division is calculated first. Then, the next smallest integer is determined: 22 / 8 = 2.75 becomes 2, while 22 / -8 = -2.75 becomes -3. If at least one of the operands is a number with decimal places, the result also appears as such a number: 7 / 2.5 = 2.8 becomes 2.0.

The modulo operator % calculates the remainder of an integer division, as in the following examples:

```
>>> 22 % 8
6
>>> 22.5 % 8.5
5.5
```

The integer division 22 // 8 results in "2 with a remainder of 6." The modulo operator returns the remainder 6. The integer division 22.5 // 8.5 results in "2.0 with a remainder of 5.5." The modulo operator returns the remainder 5.5.

2.1.4 Ranking and Parentheses

As in mathematics, the PEMDAS rule also applies in Python, which means that multiplication and division operations have priority over addition and subtraction operations. Setting parentheses means that the expressions within the parentheses will be calculated first. Let's take a look at two examples:

```
>>> 7 + 2 * 3
13
>>> (7 + 2) * 3
27
```

In the first calculation, 2 and 3 are multiplied first, then 7 is added. In the second calculation, 7 and 2 are added first and the result multiplied by 3.

2.1.5 Exercise

u_basic_arithmetic_operations

We assume that Python is installed as described earlier in Chapter 1. Open the IDLE development environment, then determine the results of the following math problems:

```
13 - 5 * 2 + 12/6
7/2 - 5/4
(12 - 5 * 2) / 4
(1/2 - 1/4 + (4 + 3)/8) * 2
```

You should get the following solutions: 5.0, 2.25, 0.5, and 2.25.

2.1.6 Variables and Assignment

Up to this point, we have used numbers for calculations only. However, if numbers are required several times in the course of a calculation, those values can be saved in variables.

Let me demonstrate this feature by converting a distance from miles to kilometers. In this context, the following applies: 1 mile = 1.609344 kilometers.

```
>>> mi = 1.609344
>>> 2 * mi
3.218688
>>> 5 * mi
8.04672
>>> 22.5 * mi
36.21024
>>> 2.35 * mi
3.7819584000000006
```

The conversion factor is initially saved in the `mi` variable. In this way, you can carry out multiple conversions in succession. The values are calculated and output in kilometers for 2 miles, 5 miles, 22.5 miles, and 2.35 miles.

The conversion factor is stored with an assignment (`mi = 1.609344`). The `mi` variable is assigned the value to the right of the equals sign.

> **Note**
>
> In Python variables, you can save integers, numbers with decimal places, character strings (i.e., texts), and other objects of a program in variables by means of assignments.
>
> Mathematically speaking, the result of the last calculation is incorrect because 2.35 miles correspond to 3.7819584 kilometers, that is, without an additional 6 at the 16th decimal place after the decimal point. Where does that incorrect display come from? In contrast to integers, numbers with decimal places cannot be stored with mathematical precision. This limitation comes into play in a few places in this book. In many calculations, however, the deviation is so small that it is negligible in practice.
>
> A value in kilometers is often rounded to three decimal places (i.e., to the nearest meter). I will discuss such formatted output later in Chapter 5, Section 5.2.2.

You can choose the name of a variable as you wish, provided you observe the following rules:

- It can consist of the letters "a" to "z," the letters "A" to "Z," any digit, or the _ character (underscore).

- It must not begin with a number.

- It must not correspond to any of the reserved words in the Python programming language. These reserved words are and, as, assert, await, break, class, continue, def, del, elif, else, except, exec, False, finally, for, from, global, if, import, in, is, lambda, None, nonlocal, not, or, pass, raise, return, True, try, while, with, and yield.

- Make sure that you enter uppercase/lowercase characters correctly. Names and statements must be written exactly as specified. The names mi and Mi denote different variables.

2.1.7   Exercise

u_inch

The following conversion applies to the linear measure "inch" with 1 inch corresponding to 2.54 centimeters. Convert the following values into centimeters: 5 inches, 20 inches, and 92.7 inches. Simplify your calculations by first saving the conversion factor in a variable.

As mentioned earlier, in is a reserved keyword and is not suitable as a variable name. You should get the following solutions: 12.7, 50.8, and 235.458.

2.2   Our First Program

Let's now enter, save, and call our first Python program. I will explain this process in detail. Later, you need to carry out the individual steps for each Python program.

All programs and explanations initially refer to Python on Windows. If differences with Ubuntu Linux or macOS exist, I will mention them in the relevant section. You'll also find all the programs in the materials accompanying the book at *https://rheinwerk-computing.com/5876*.

2.2.1   Hello World

The output of the first Python program is:

```
Hello world
```

The program displays the "Hello world" text on the screen—often the first program you write when learning any new programming language.

2.2.2 Entering a Program

To enter the program, in IDLE, you first need to call the **New File** command in the **File** menu. Then, a new window named **untitled** opens, while the main **IDLE Shell 3.12.2** window fades into the background.

In the new window, enter the program shown in Figure 2.2.

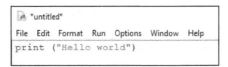

Figure 2.2 Entering the Program in a New Window

The built-in print() function can output character strings (i.e., texts) to the screen, among other things. Strings are placed within double quotation marks.

2.3 Saving and Running

To see the result of the program, you must first save it in a file and then run it.

2.3.1 Saving

To save the program, select the **Save** command from the **File** menu in the current untitled window. The program should be saved in the *C:\Python\Examples* directory, in a file named *hello.py*, as shown in Figure 2.3.

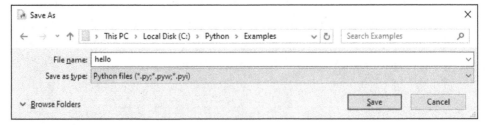

Figure 2.3 Saving the Python Program

After clicking the **Save** button, the saving process is finished. You can freely choose a name for the file (in our case, *hello*). The *.py* file extension is mandatory for Python programs and is automatically added.

The window with the program is shown in Figure 2.4. The filename or the complete path is shown in the title bar at the top.

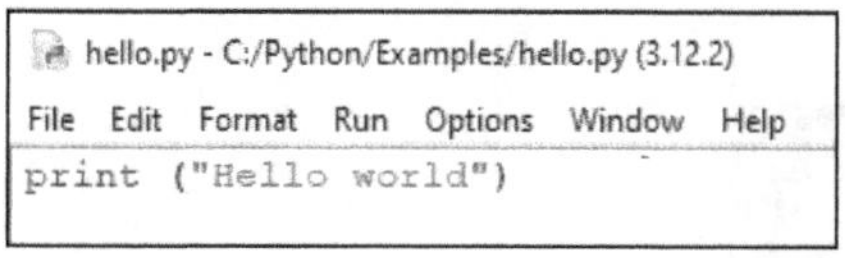

```
hello.py - C:/Python/Examples/hello.py (3.12.2)
File  Edit  Format  Run  Options  Window  Help
print ("Hello world")
```

Figure 2.4 Filename and Path in the Title Bar

2.3.2   Running on Windows

On Windows, you can launch the program in two different ways:

1. Within the IDLE development environment (we use this approach for most of our programs in this book)

2. From the command line

Within the IDLE Development Environment

In the window that contains the IDLE program, select the **Run Module** menu command in the **Run** menu or press F5 . The main window of the development environment, the **IDLE Shell**, returns to the foreground and the output text is displayed, as shown in Figure 2.5.

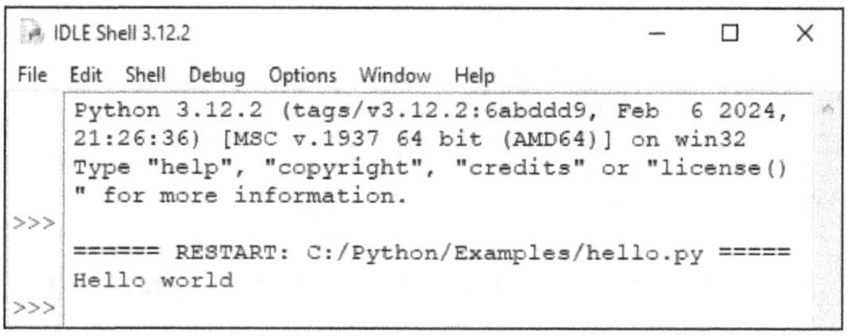

```
IDLE Shell 3.12.2                                     —    □    ×
File  Edit  Shell  Debug  Options  Window  Help
    Python 3.12.2 (tags/v3.12.2:6abddd9, Feb  6 2024,
    21:26:36) [MSC v.1937 64 bit (AMD64)] on win32
    Type "help", "copyright", "credits" or "license()
    " for more information.
>>>
    ====== RESTART: C:/Python/Examples/hello.py ======
    Hello world
>>>
```

Figure 2.5 Result of the Program

You can then use the taskbar to bring the program window back to the foreground.

From the Command Line

The program in the *hello.py* file is a command line program, which is why it only generates a simple output on the screen. In Chapter 10 and in Chapter 11, you'll learn how to create programs with graphical user interfaces (GUIs) using Python.

To call a program from the command line, you must first switch to the command line level on Windows. For this step, execute the **Command Prompt** program in the **Windows System** group in the start menu.

Then, a command line window is displayed, as shown in Figure 2.6.

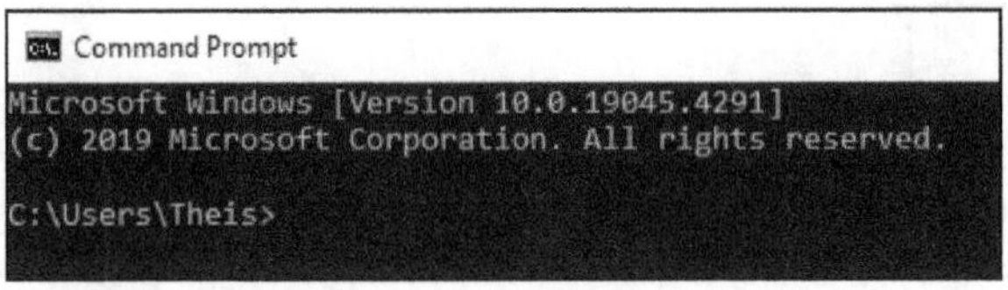

Figure 2.6 Command Line Window

Enter the command `cd \Python\Examples` to change to the *C:\Python\Examples* directory, where the program is saved in the *hello.py* file. Note the space after the `cd` command.

You can initiate the execution of the program by calling the Python interpreter via the
`python hello.py` statement. Then, the output shown in Figure 2.7 appears.

You can leave this command line window open for later calls of Python programs or
close it by entering the `exit` command.

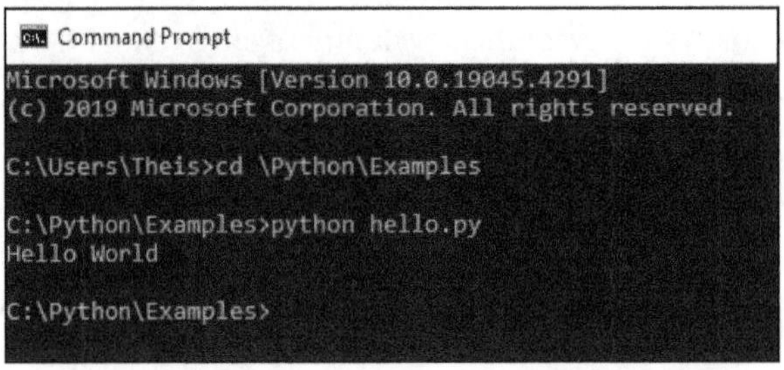

Figure 2.7 Call in the Command Line Window

Because you selected **Add Python to Path** during the installation process in Chapter 1,
Section 1.5, the Python interpreter is available in every directory.

2.3.3   Running on Ubuntu Linux and on macOS

You can also call the program on Ubuntu Linux and on macOS in two ways: first within
the IDLE development environment (see our earlier discussion in Section 2.3.2) and second from a terminal.

Open a terminal and go to the directory in which you have saved the *hello.py* file. On
Ubuntu Linux, you'll find the terminal after entering the term in the **dash**. On macOS,
you can access the terminal via the **dock**. Open the **Launchpad** and select the **Other**
group.

Initiate the execution of the program for Python 3 using the `python3 hello.py` statement, as shown in Figure 2.8 and Figure 2.9.

Figure 2.8 Calling Python 3 on Ubuntu Linux

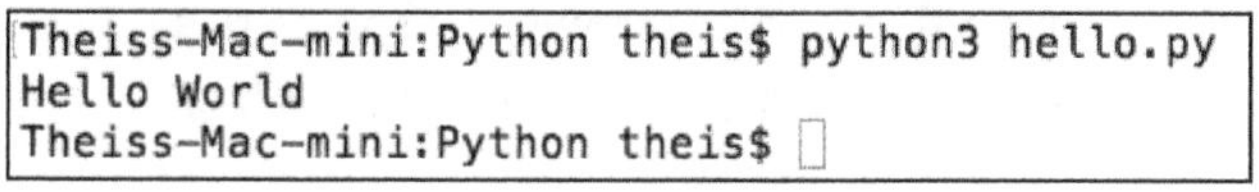

```
Theiss-Mac-mini:Python theis$ python3 hello.py
Hello World
Theiss-Mac-mini:Python theis$ 
```

Figure 2.9 Calling Python 3 on macOS

By entering `python3 -V` (uppercase "V"), you can also determine the exact Python version number at this point.

> **Note**
>
> You can find more command line commands for Ubuntu Linux and macOS in Appendix A, Section A.4.

2.3.4   Comments

For more extensive programs, you should add comments to the program text for explanation. Single-line comments are introduced by the hash character (#) and extend to the end of the line. Multiline comments start and end with three double quotation marks. The system does not regard comments as program steps, which is why they are not executed. In the following listing, our first program is supplemented by comments.

```
# My first program
print("Hello world")        # An output
"""Comment in
   multiple lines"""
```

Listing 2.1 hello.py File with Comments

2.3.5   Concatenating Outputs

The `print()` function can also be used to produce multiple outputs in one line. The program is changed further.

```
# My first program
print("Hello", "world")
```

Listing 2.2 hello.py File with Multiple Outputs

The individual parts of the output are separated by commas. A space is automatically inserted after each part of the output. If you want to influence this behavior, see Chapter 5, Section 5.2.1, for more information about the print() function.

2.3.6   Long Outputs

Character strings that are output via the print() function can become exceedingly long, possibly even extending beyond the right margin within the editor, which can make the program code confusing. To avoid this problem, you can spread this type of character strings across multiple lines—this procedure is also used in this book due to the limits of the printed page.

Although it is not necessary for this short program, we'll now spread the code across several lines to make it clearer:

```
# My first program
print("Hello",
      "world")
print("Hello "
      "world")
```

Listing 2.3 The hello.py File with Long Outputs

A good idea is to insert the line break after a comma, as in the first output of "Hello world." As explained earlier, a space is automatically inserted.

Individual long texts can be broken down into sub-texts, as in the second output of "Hello world." Sub-texts do not need to be separated by commas but must be delimited using quotation marks. You must now manually set the separating space between "Hello" and "world."

Note

In Chapter 5, Section 5.1.2, I explain how you can generally split up long program lines.

Chapter 3
Programming Course

The following programming course with detailed explanations will introduce you to programming using Python, step by step.

The course is accompanied by a programming project in which you develop a game—an entertaining way to get to know the Python language. This programming project combines many individual parts into a complete whole.

3.1 Programming a Game

Our game will be continuously expanded and improved over the course of this book. To start, I want to describe the course of the game.

After starting the program, a calculation task is set. The mentally calculated result is entered as a suggested solution, and the program evaluates the input. Let's look at how this game is played:

```
The task: 9 + 26
Please enter a suggested solution:
34
34 is wrong
Please enter a suggested solution:
35
35 is correct
Result:  35
Number of attempts: 2
```

The game is created in several individual steps. A simple version is created first. As your knowledge increases, we'll develop increasingly complex versions. You can use the skills you learn in each section directly to improve the gameplay.

In this way, this game accompanies you through the entire book. Among other things, it will be expanded to include the following options:

- Multiple tasks will be set.
- The time required is measured.
- The name of the player and the time it took will be recorded in a highscore list, which is stored permanently in a file or database.

- The highscore list will be updated with new results and displayed on the screen.
- A version of the game will feature a graphical user interface (GUI).

3.2 Variables and Operators

Variables are required to store values. Operators are used to perform calculations.

3.2.1 Assignment, Calculation, and Output

The following program performs a simple calculation using an operator. The result of the calculation is assigned to a variable using the equals sign. This calculation step is followed by an output of the result, as described in Chapter 2, Section 2.1.6.

```
a = 5
b = 3
c = a + b

print("The task:", a, "+", b)
print("The result:", c)
```

Listing 3.1 assignment.py File

The output of this program reads:

```
The task: 5 + 3
The result: 8
```

A value is saved in each of the two variables: a and b. The two values are added together, and the result is assigned to variable c. The task is output, followed by the result. When you use this program, you cannot yet intervene in the process.

> **Note**
> Since the a, b, and c variables have been assigned integers, they are currently variables for integers. However, Python is a dynamically typed language. For this reason, variables can change their types as a result of another assignment during the course of the program.

3.2.2 String Literals

String literals are used for the convenient embedding of simple variables but also complex expressions in character strings (i.e., in texts). Let's now extend our program from the previous section into a version with string literals.

```
a = 5
b = 3
c = a + b

print(f"The task: {a} + {b}")
print(f"The result: {c}")
print(f"The result without saving: {a + b}")
print(f"Another calculation: {(12 - 5 * 2) / 4}")
```

Listing 3.2 literal.py File

The output of the program reads:

```
The task: 5 + 3
The result: 8
The result without saving: 8
Another calculation: 0.5
```

The f character before the start of the string introduces a string literal. You can embed multiple expressions in the character string, each one enclosed by curly brackets. An expression can contain operators, parentheses, and the names of variables, among other things. The value of an expression is determined and output together with the rest of the character string. There must not be any space between the f character and the quotation mark.

Chapter 5, Section 5.2.2, contains more information on the capabilities of string literals.

3.2.3   Entering a String

This section introduces the built-in input() function, which we'll use to enter a character string (i.e., a text). Here's a small sample program:

```
print("Please enter a text")
x = input()
print("Your input:", x)
```

Listing 3.3 input_text.py File

The output could look as follows:

```
Please enter a text
Python
Your input: Python
```

A text is entered. This text is saved in the x variable and then output.

Note

As the x variable has been assigned a string, it is currently a variable for *strings*.

Tip

On international keyboards, you can access characters like the curly brackets { and } using the AltGr key to the right of the space bar.

3.2.4 Entering a Number

As the game progresses, let's say we need to continue using the text entered as a number. For this goal, the character string returned by the input() function must be converted into an integer.

The following functions are available for conversion, among others:

- The built-in int() function converts a character string containing a valid integer into an integer. If the string does not contain a valid integer, the program terminates.

- The built-in float() function converts a character string containing a valid number into a number with decimal places. If the string does not contain a valid number, the program terminates. Only a point is allowed as a decimal separator.

In Section 3.6, you'll learn how you can intercept the termination of a program. Until we reach that section, I will simply assume that correct entries have been made.

Here's an example that contains the functions int() and float():

```
print("Please enter an integer")
a = input()
print("Your input:", a)

b = int(a)
print("As integer:", b)
print("Doubled:", b * 2)

print("Please enter a number")
x = input()
print("Your input:", x)

y = float(x)
print("As number:", y)
print("Doubled:", y * 2)
```

Listing 3.4 input_number.py File

The output could look as follows:

```
Please enter an integer
3
Your input: 3
As integer: 3
Doubled: 6
Please enter a number
3.5
Your input: 3.5
As number: 3.5
Doubled: 7.0
```

As the player/user, we enter the first character string. This string is then converted into an integer using the built-in `int()` function. The number and the double of the number are output.

Then, a second character string is entered, which is converted into a number with decimal places using the built-in `float()` function. The number and the double of the number are output.

3.2.5   Our Game: Version with Input

Our game's program, in which a mental math problem is to be solved, receives an input. The program will be changed in the following way:

```
a = 5
b = 3
c = a + b
print(f"The task: {a} + {b}")

print("Please enter a suggested solution:")
z = input()
number = int(z)

print("Your input:", z)
print("The result:", c)
```

Listing 3.5 game_input.py File

A possible output of the program could read as follows:

```
The task: 5 + 3
Please enter a suggested solution:
9
```

```
Your input: 9
The result: 8
```

The program displays the "Please enter a suggested solution:" prompt and stops. The text you now enter is saved in the z variable. The z string is converted into an integer using the int() function.

3.2.6   Exercises

---

u_input_inch

Write a program for entering and converting any inch values into centimeters. The conversion factor should be saved in a variable as in exercise u_inch from Chapter 2, Section 2.1.7. Save the program in the *u_input_inch.py* file. Call the program and test it. An example output might resemble the following:

```
Please enter the inch value:
3.5
3.5 inches are 8.89 cm
```

---

u_input_salary

Write a program in the *u_input_salary.py* file to simplify the calculation of taxes. After starting the program, you'll be asked to enter your monthly salary. Then, 18% of this amount is calculated and output. An example output might resemble the following:

```
Enter your salary in $:
2500
This results in a tax of 450.0 $
```

3.2.7   Random Numbers

The random module in Python provides a generator for creating random numbers, as they are often required for games and simulations. In our game, we'll use it to create the input for the mental calculation task. The functions of the random generator are located in an additional module that must first be imported.

The program is now changed in the following way:

```
import random
random.seed()

a = random.randint(1,10)
b = random.randint(1,10)
```

```
c = a + b
print(f"The task: {a} + {b}")

print("Please enter a suggested solution:")
number = int(input())
print("Your input:", number)
print("The result:", c)
```

Listing 3.6 game_random_number.py File

A possible output of the program, depending on the random values provided, looks as follows:

```
The task: 8 + 3
Please enter a suggested solution:
7
Your input: 7
The result: 11
```

One of the strengths of Python is its many additional modules, each of which contains numerous additional functions. You can easily integrate new modules into your programs using the import statement. Then, you can call the functions found in these modules using the module_name.function_name notation.

Calling the seed() function of the random module causes the random number generator to be initialized with the current system time. Otherwise, instead of a random selection, what might occur is that the same numbers would be provided again and again.

The randint() function of the random module returns a random integer in the specified range, in this case, a random number between 1 and 10, inclusive.

In the number = int(input()) statement, two nested functions are called: first the "inner" input() function so that the program accepts an input and then the "outer" int() function so that the input is converted into an integer.

> **Note**
>
> The functions of the random module can be used for the random values of a game. If you need random values for the purpose of encryption or for security reasons, you should use the functions of the secrets module, which has been available since Python 3.6 (see Chapter 5, Section 5.4).

3.2.8 Determining a Type

The type() function outputs the type (also called the "class" or the "data type") of an object, not only for number types and character strings.

Let's take a look at a programming example:

```
a = 2
print("2:", type(a))

b = 12/6
print("12/6:", type(b))

c = 12//6
print("12//6:", type(c))

d = "abc"
print("abc:", type(d))
```

Listing 3.7 data_type.py File

This program generates the following output:

```
2: <class 'int'>
12/6: <class 'float'>
12//6: <class 'int'>
abc: <class 'str'>
```

The a variable contains the value 2, which means that it is currently a variable for an integer. The int type is available for this purpose.

The result of a mathematical division is a number with decimal places. The b variable is therefore currently a variable for a number with decimal places. The float type is available for this purpose.

The c variable contains the result of an integer division and is therefore currently a variable of the int type.

A string was assigned to the d variable. It is therefore currently a variable of the str type (short for *string*).

3.3   Branches

In our programs so far, all statements were executed in sequence. However, to control a program's flow, you often need to use branches. In this context, a comparison is used to decide which branch of the program should be executed.

3.3.1   Comparison Operators

A comparison is formulated by means of a comparison operator. Table 3.1 lists the comparison operators and their meanings.

Operator	Meaning
>	Greater than
<	Less than
>=	Greater than or equal to
<=	Less than or equal to
==	Equal to
!=	Not equal to

Table 3.1 Comparison Operators

A comparison returns one of the two truth values (also referred to as *Boolean values*): True or False. Truth values can be saved in a variable for later use, which then becomes a variable of the bool type. Let's consider an example.

```
x = 12
print("x:", x)

print("x == 12:", x == 12)
print(type(x == 12))

z = (x != 12)
print("x != 12:", z)
print(type(z))
```

Listing 3.8 operator_comparison.py File

The output of the program reads:

```
x: 12
x == 12: True
<class 'bool'>
x != 12: False
<class 'bool'>
```

The x variable has the value 12. A comparison is a *Boolean expression*. The result of the first comparison (x == 12) is output immediately along with the type of the expression. The result of the second comparison (x != 12) is temporarily saved in the z variable. This variable is output together with its current type. The parentheses around the comparison are not necessary, as the assignment only takes place after the comparison. They have been written for clarification purposes only.

Truth values are usually used to control branches or loops. Chapter 4, Section 4.7, contains more information on truth values.

3.3.2   Branch Using if

Our next example checks whether a randomly determined integer is positive. If so, the output will read "This number is positive"; otherwise, the output will read "This number is 0 or negative." Thus, either one or the other will be output.

```
import random
random.seed
x = random.randint(-5, 5)
print("x:", x)

if x > 0:
    print("This number is positive")
else:
    print("This number is 0 or negative")
```

Listing 3.9 branch_if.py File

A possible output of this program could read as follows:

```
x: 3
This number is positive
```

First, the x variable is assigned a random value, in this case, 3.

A branch is initiated using if. Then, a condition is formulated (e.g., x > 0), which results in either *True* or *False*. This condition is followed by a colon.

Then follow one or more statements that are only executed if the condition is *True*. The statements must be indented within the if branch by pressing [Tab] so that Python can recognize the branch to which they belong.

You can identify the alternative part to a branch that is introduced using the else statement, which is again followed by a colon. One or more statements are then noted, which are only executed if the condition is *False*. These statements must also be indented.

> **Note**
>
> The Integrated Development and Learning Environment (IDLE) uses indentation by default, which means that a program is already clearly structured at the development stage. This behavior applies not only to branches, but also to loops and other control structures.

3.3.3 Our Game: Version with Input Evaluation

In our program, a game in which a mental math problem is to be solved, is extended by an evaluation of the user's input. For this purpose, a simple branch checks whether the correct solution has been entered. Thus, the program changes in the following ways:

```
import random
random.seed()


a = random.randint(1,10)
b = random.randint(1,10)
c = a + b
print(f"The task: {a} + {b}")

print("Please enter a suggested solution:")
number = int(input())

if number == c:
    print(number, "is correct")
else:
    print(number, "is wrong")
    print("Result:", c)
```

Listing 3.10 game_branch.py File

A possible output of the program could read as follows:

```
The task: 2 + 8
Please enter a suggested solution:
11
11 is wrong
Result: 10
```

In this version, the input is converted into a number. If this number corresponds to the result of the calculation, the text "... is correct" will appear. If the input does not correspond to the calculated result, the text "... is wrong" will appear, together with the correct result.

3.3.4 Exercise

u_branch_if

Our simplified program for calculating the tax needs to be changed. After starting the program, you're supposed to enter your monthly salary. If this value is over $2,500, you owe 22% tax; otherwise, 18%. Use the *u_branch_if.py* file for this exercise.

Only one entry is required. Within the program, the tax rate to be applied is to be decided on the basis of the salary. An example output might resemble the following:

```
Enter your salary in $:
3000
This results in a tax of 660.0 $
```

Or perhaps the following:

```
Enter your salary in $:
2000
This results in a tax of 360.0 $
```

3.3.5   Multiple Branches

In many applications, you have more than two options to choose from. To make this work, multiple branches are required.

Our next example checks whether a random number is positive, negative, or equal to 0, for which a corresponding message will then be output.

```python
import random
random.seed
x = random.randint(-5, 5)
print("x:", x)

if x > 0:
    print("x is positive")
elif x < 0:
    print("x is negative")
else:
    print("x is equal to 0")
```

Listing 3.11 branch_multiple.py File

A possible output of this program could read as follows:

```
x: -3
x is negative
```

The branch is initiated using if. If x is positive, the following indented statements will be executed.

A further condition is formulated after elif. It is only checked if the first condition (after the if) does not apply. If x is negative, the indented statements that follow will be executed.

The statements that follow the else are only executed if none of the two previous conditions apply (after if and after elif). In this case, x must be 0 because it is neither positive nor negative.

> **Note**
> Within a branch, you can have multiple elif statements, which are checked one after the other until the program arrives at a condition that applies. The remaining elif statements or an else statement are no longer taken into account in this case. If no condition applies, the statements are executed after the else statement.

3.3.6 Exercise

> **u_branch_multiple**
> Now, we want to modify our program for calculating taxes further (*u_branch_multiple.py* file). After entering the monthly salary, the tax should be calculated according to the following table:
>
> | Salary | Tax Rate |
> |---|---|
> | More than $4,000 | 26% |
> | From $2,500 to $4,000 | 22% |
> | Less than $2,500 | 18% |

3.3.7 Conditional Expression

If only one value is assigned to a branch in all cases, you can also use a *conditional expression*. The result of the expression can be saved or output immediately, as in our next program.

```
import random
random.seed
x = random.randint(-3, 3)
print("x:", x)

output = "positive" if x>0 else "0 or negative"
print("This number is", output)

print("This number is", "positive" if x>0 else "0 or negative")

print("This number is", "positive" if x>0 else "negative" if x<0 else "equal to 0")
```

Listing 3.12 conditional_expression.py File

A possible output of the program could read as follows:

```
x: -3
This number is 0 or negative
This number is 0 or negative
This number is negative
```

The first conditional expression is used to assign the "positive" value to the output variable if x is greater than 0; otherwise, the value "0 or negative." The assigned value of output is then output.

The result of the second conditional expression is output immediately.

Conditional expressions can be nested to form multiple branches. The third conditional expression enables three different outputs.

3.3.8   Logical Operators

The logical operators and, or, and not can be used to link several conditions together.

- A condition that consists of one or more individual conditions, each of which is linked with the and operator, results in *True* if *each* of the individual conditions results in *True*.
- A condition that consists of one or more individual conditions which are linked with the or operator is *True* if *at least one* of the individual conditions is *True*.
- The not operator reverses the truth value of a condition; that is, a false condition becomes true, and a true condition becomes false.

This correlation is also shown in Table 3.2 and Table 3.3.

Condition 1	Condition 2	Result for AND	Result for OR
True	True	True	True
True	False	False	True
False	True	False	True
False	False	False	False

Table 3.2 Effect of the Logical Operators AND and OR

Condition	Result
True	False
False	True

Table 3.3 Effect of the Logical Operator NOT

For example:

```
x = 12
y = 15
z = 20
print(f"x:{x}, y:{y}, z:{z}")

if x<y and x<z:
    print("x is the smallest number")

if y>x or y>z:
    print("y is not the smallest number")

if not y<x:
    print("y is not smaller than x")
```

Listing 3.13 operator_logical.py File

The output of the program reads:

```
x:12, y:15, z:20
x is the smallest number
y is not the smallest number
y is not smaller than x
```

Condition 1 is *True* if x is less than y *and* less than z. This rule applies to the given initial values. Condition 2 is *True* if y is greater than x *or* y is greater than z. As the first condition is true, the entire condition is true: y is not the smallest of the three numbers.

Condition 3 is true if y is *not* smaller than x (as in our case). You can also have an else branch for all branches.

3.3.9 Exercises

u_operator

We need to change our program for calculating taxes again (*u_operator.py* file). The table now looks as follows:

Salary	Marital Status	Tax Rate
> 4,000 $	Single	26%
> 4,000 $	Married	22%
<= 4,000 $	Single	22%
<= 4,000 $	Married	18%

---

u_date

Develop a program to check a date (*u_date.py* file). After calling the program, the three components of a date should be entered individually. The system then determines whether these inputs constitute a valid date. During development and testing, your program should follow these steps:

1. Examine the entered value for the day. If this value is less than 1 or greater than 31, the whole input is an invalid date.

2. Examine the entered value for the month. If this value is less than 1 or greater than 12, the whole input is an invalid date.

3. Output the value of the last day of the month in question. Remember that there are only three possible cases: 28, 30, or 31 days. The rules for leap years are not yet observed.

4. Examine the entered value for the day. Specify whether this value is less than 1 or greater than the last day of the month in question.

5. Examine the entered value for the year. Indicate whether the year is a leap year. The simplified rule for a leap year is: If the value can be divided by 4 without a remainder, it is a leap year.

6. Combine the previous steps with each other. If the value for the day is less than 1 or greater than the last day of the month in question (taking into account the rule for a leap year), it is an invalid date; otherwise, the whole input is a valid date.

Let's expand the program. The complete rule for a leap year is: If the value can be divided by 4 without a remainder, but not by 100 without a remainder, it is a leap year. However, it is also a leap year if the value can be divided by 400 without a remainder.

Executing our finished program could result in the following output:

```
Enter the day of the date:
29
Enter the month of the date:
2
Enter the year of the date:
2000
Last day: 29
Valid date
```

3.3.10   Multiple Comparison Operators

Conditions can also contain multiple comparison operators. Some branches are easier to understand this way. Here's an example:

```
x = 12
y = 15
z = 20
print(f"x:{x}, y:{y}, z:{z}")

if x < y < z:
    print("y is between x and z")
```

Listing 3.14 operator_multiple.py File

The output of the program reads:

```
x:12, y:15, z:20
y is between x and z
```

The x < y < z condition corresponds to the linked x < y and y < z condition. However, it is easier to read in the short form.

3.3.11 Our Game: Version with Exact Evaluation of the Input

The input for our program, a game in which a mental calculation task is to be solved, can evaluate inputs more precisely in various ways:

- Using multiple branches
- Using logical operators
- Using conditions with multiple comparison operators

Let's now change our program in the following way:

```
import random
random.seed()

a = random.randint(1,10)
b = random.randint(1,10)
c = a + b
print(f"The task: {a} + {b}")

print("Please enter a suggested solution:")
number = int(input())

if number == c:
    print(number, "is correct")
elif number < 0 or number > 100:
    print(number, "is far off")
elif c-1 <= number <= c+1:
    print(number, "is close")
```

```
else:
    print(number, "is wrong")

print("Result:", c)
```

Listing 3.15 game_operator.py File

A possible output of the program could read as follows:

```
The task: 7 + 8
Please enter a suggested solution:
16
16 is close
Result: 15
```

A total of four options are offered: via if, twice elif, and else. If the input is less than 0 *or* greater than 100, the input is far from the correct result. This question is resolved using the logical operator or.

If the entered number differs from the correct result by only the value 1, the entry is close. This closeness is determined using a condition that contains several comparison operators.

3.3.12   Branch with match

Since Python 3.10, match has been another option for creating multiple branches in a clear form. Let's consider a simple example.

```
import random
random.seed()

x = "Paris"
print("Value =", x)
match x:
    case "Paris":
        print("France")
    case "Rome":
        print("Italy")
    case "Madrid":
        print("Spain")
    case _:
        print("Unknown country")
print()
...
```

Listing 3.16 branch_match.py File, Part 1 of 3

The output of this program reads:

```
Value = Paris
France
```

The random generator is not required until later in the program.

A string is saved in the x variable. The match keyword is followed by the expression to be compared with. This expression can be a variable or the result of a calculation. The colon is followed by the different cases which are indented.

A single case begins with the case keyword. This keyword is followed by the value with which the match expression is compared, followed by a colon. Then follow the statements, again indented, if the comparison in question is successful.

The _ character can be used to formulate another case at the end. If none of the above cases apply, the statements for this default case are executed.

This example is focused on the comparison of character strings. However, you can also compare individual characters or integers.

You can also compare figures with decimal places, but these values are not saved with mathematical precision. For this reason, even a small deviation could lead to an incorrect result.

Let's look at the second part of this matching program:

```
...
x = random.randint(1,6)
print("Value =", x)
match x:
    case 1 | 3 | 5:
        print("odd")
    case 2 | 4 | 6:
        print("even")
    case _:
        print("No dice value")
print()
...
```

Listing 3.17 branch_match.py File, Part 2 of 3

A possible output of this part of the program reads as follows:

```
Value = 2
even
```

The random generator is used to determine a random integer between 1 and 6 (i.e., a value you can roll on a single die). The | operator enables the linking of multiple cases. If a 2, 4, or 6 is rolled, the text "even" is displayed.

Now, we've come to the third and final part of this program:

```
...
x = random.randint(1,10)
print("Value =", x * 1.5)
match x * 1.5:
    case x if x < 5:
        print("small value")
    case x if x > 11:
        print("large value")
    case _:
        print("medium value")
```

Listing 3.18 branch_match.py File, Part 3 of 3

A possible output of this part of the program reads as follows:

```
Value = 9.0
medium value
```

An embedded if enables you to use comparison operators. The case keyword is followed by a comparison of the examined expression. As soon as the first comparison is correct, the corresponding statements will be executed.

In this way, you can compare numbers that have decimal places. In the first comparison, the possible value is 4.5 below the limit, and the next possible value is 6.0 above the limit.

3.3.13 Order of Operations

Many expressions contain several operators. So far, we've used calculation operators, comparison operators, and logical operators. The order in which the operators are executed is important. The sub-steps in which higher-ranking operators are involved are executed first.

Table 3.4 indicates the order of precedence of the operators we've used so far in Python, starting with the operators that have the highest precedence. Operators of equal rank are included together on one line. Sub-steps containing multiple operators of the same rank are executed from left to right.

Operator	Meaning
+ -	Positive sign of a number, negative sign of a number
* / % //	Multiplication, division, modulo, integer division

Table 3.4 Ranking of the Operators Used So Far

Operator	Meaning
+ -	Addition, subtraction
< <= > >= == !=	Less, less than or equal to, greater, greater than or equal to, equal to, not equal to
not	Logical NOT
and	Logical AND
or	Logical OR

Table 3.4 Ranking of the Operators Used So Far (Cont.)

3.4 Loops

In addition to branches, another important structure is used for controlling programs: the loop. It enables the repeated execution of program steps. In this context, a distinction is made between two types of loops: the for loop and the while loop. The respective area of application of the two types is defined by the following characteristics:

- A for loop is used if a program step is supposed to be executed repeatedly for a regular and known sequence of values.
- A while loop is used if it only becomes clear after the program has been called whether a program step is to be executed and how often it is to be repeated.

A for loop is also referred to as a *counting loop*; a while loop, as a *conditional loop*.

3.4.1 Loops with for

In a for loop, all elements of an *iterable object* (in short, an *iterable*) are run through. An iterable is an object that consists of several elements and can be run through, such as a list of several numbers or a list of multiple character strings.

A character string itself can also be iterated, as its individual characters can be run through as elements of the string. You'll learn about other iterables later in the book.

A copy of the elements is used within the loop. A change to the copy has no retroactive effect on the original.

Let's consider some examples:

```
for number in 8, 3, 7:
    print(f"Number: {number}, Square: {number * number}")
print()

for city in "Paris", "Rome", "Madrid":
    print(f"City: {city}")
```

```
print()

for characters in "Rome":
    print(f"Characters: {characters}")
print()

a = 2
b = 8
print(f"a:{a}, b:{b}")
for i in a, b:
    i = i + 1
    print(f"Copy: {i}")
print(f"a:{a}, b:{b}")
```

Listing 3.19 loop_for.py File

The following output is generated:

```
Number: 8, Square: 64
Number: 3, Square: 9
Number: 7, Square: 49

City: Paris
City: Rome
City: Madrid

Characters: R
Characters: o
Characters: m
Characters: e

a:2, b:8
Copy: 3
Copy: 9
a:2, b:8
```

The notation for [Variable] in [Iterable]: means that the entire iterable is run through and the variable takes on each value of the iterable in turn.

In the first for loop, a sequence of numbers is run through. Within the loop, each of these numbers can be accessed via the number variable and is output together with its square.

In the second for loop, a sequence of character strings is run through. Within the loop, each of these character strings can be accessed via the city variable and is output together with a text.

In the third for loop, a single character string is run through. Character strings consist of a sequence of individual characters. Within the loop, each of these characters can be accessed via the characters variable and is output together with a text.

In the fourth for loop, a sequence of variables is run through. Within the loop, a copy of the current variable can be accessed via the i variable. The copy is modified and output, which does not change the values of the original variables.

As is the case with a branch using if, you must use a colon and indent the statements in the loop.

3.4.2 Terminating a Loop Using break

The break keyword causes a loop to terminate immediately. Such a termination is usually linked to a condition and used in special cases. Here's an example:

```
for i in 12, -4, 20, 7:
    if i*i > 200:
        break
    print(f"Number: {i}, Square: {i*i}")
```

Listing 3.20 loop_break.py File

The following output is generated:

```
Number: 12, Square: 144
Number: -4, Square: 16
```

If the square of the current number is greater than 200, the loop is exited immediately. The output within the loop will then also no longer take place.

The break keyword can be used in both for loops and while loops.

3.4.3 Continuing a Loop Using continue

The continue keyword is used to immediately terminate the current run of a loop. The program then continues with the next run of the loop. Let's take a look at the following program:

```
for i in 1, 2, 3, 4, 5, 6:
    print("Number:", i)
    if 3 <= i <= 5:
        continue
    print("Square:", i*i)
```

Listing 3.21 loop_continue.py File

The output of this program reads:

```
Number: 1
Square: 1
Number: 2
Square: 4
Number: 3
Number: 4
Number: 5
Number: 6
Square: 36
```

The loop runs through the numbers from 1 to 6 and all these figures are output. If the current number is between 3 and 5, the rest of the loop is skipped, and the next loop pass is started immediately. Otherwise, the square of the number will be output.

3.4.4 Nested Control Structures

As the last two programs have shown, control structures (i.e., branches and loops) can be nested. Therefore, a control structure contains another control structure, which in turn can contain a control structure, and so on.

To illustrate this concept, let's look at another example:

```
for x in -2, -1, 0, 1, 2:
    if x > 0:
        print(x, "positive")
    else:
        if x < 0:
            print(x, "negative")
        else:
            print(x, "equal to 0")
```

Listing 3.22 nesting.py File

The following output is generated:

```
-2 negative
-1 negative
0 equal to 0
1 positive
2 positive
```

The outermost control structure is a for loop. All single indented statements are executed several times in accordance with the loop control.

The first branch is initiated with the outer if statement. If x is greater than 0, the following double indented statement is executed. If x is not greater than 0, the inner if statement is examined after the outer else statement.

If the condition of the inner `if` statement is met, the subsequent triple indented statements are executed. If it does not apply, the triple indented statements that follow the inner `else` statement are executed.

You should pay particular attention to the multiple indentations so that the depth of the control structure can be correctly recognized by Python.

> **Note**
>
> A colon after the header of a control structure is automatically indented with a tab jump in the IDLE development environment. This jump creates four spaces by default, making the control structures clearly recognizable.
>
> If you use a different development environment, you must ensure that at least one space is indented so that Python recognizes the control structure. It makes more sense to indent by two or more spaces so that the structure is clearly recognizable.

3.4.5 Our Game: Version with for Loop and Abort

The `for` loop is now used to run through the input and the evaluation of the mental calculation game a total of four times. This loop gives you four attempts to enter the correct result. The `break` statement is used to terminate the loop as soon as the correct result has been entered.

```python
import random
random.seed()

a = random.randint(1,10)
b = random.randint(1,10)
c = a + b
print(f"The task: {a} + {b}")

for i in 1, 2, 3, 4:
    print("Please enter a suggested solution:")
    number = int(input())
    if number == c:
        print(number, "is correct")
        break
    else:
        print(number, "is wrong")

print("Result:", c)
```

Listing 3.23 game_for.py File

The following output is generated:

```
The task: 6 + 8
Please enter a suggested solution:
13
13 is wrong
Please enter a suggested solution:
14
14 is correct
Result: 14
```

The task is determined and set once. You'll then be asked to enter a result a maximum of four times. Each entry is evaluated. If one of the first three attempts is correct, the loop will terminate prematurely.

3.4.6   Loops with for and range()

Loops are usually used for regular sequences of numbers. The use of the built-in range() function proves quite useful in this context. Let's look at an example:

```
for i in range(3,11,2):
    print(f"Number: {i}, Square: {i*i}")
```

Listing 3.24 range_three.py File

The following output is generated:

```
Number: 3, Square: 9
Number: 5, Square: 25
Number: 7, Square: 49
Number: 9, Square: 81
```

A maximum of three integers, separated by commas, can be entered within the parentheses after range:

- The first integer (in our case, 3) indicates the start of the range for which the following statements are executed.
- The second integer (in our case, 11) indicates the end of the range. It is the first number for which the statements will *no* longer be executed.
- The third integer (in our case, 2) specifies the step size or increment for the loop. The numbers for which the statements are executed are therefore spaced +2 apart.

Calling the range() function with the numbers 3, 11, and 2 thus produces the sequence: 3, 5, 7, and 9. If the range() function is only called with two numbers, an increment of 1 is assumed, as in our next example.

```
for i in range(5,9):
    print("Number:", i)
```

Listing 3.25 range_two.py File

The following output is generated:

```
Number: 5
Number: 6
Number: 7
Number: 8
```

If the range() function is only called with one number, the number of runs is fixed. The lower limit is 0; the upper limit is the specified number −1. An increment of 1 is assumed in our next example.

```
for i in range(3):
    print("Number:", i)
```

Listing 3.26 range_one.py File

Here's the output:

```
Number: 0
Number: 1
Number: 2
```

> **Note**
>
> With the range() function, one or more of the three numbers can also be negative. You should make sure to use number combinations that make sense. The range(3,-11,2) specification, for example, does not make sense, as you cannot get from the number +3 to the number -11 in steps of +2. Python intercepts such loops and does not allow them to be executed. The same applies to the range(3,11,-2) combination.

For the regular sequence of numbers with decimal places, the loop variable should be converted appropriately, for example, in the following way:

```
for x in range(18,22):
    print(x/10)
print()

x = 1.8
for i in range(4):
    print(x)
    x = x + 0.1
```

Listing 3.27 range_decimal.py File

The output is:

```
1.8
1.9
2.0
2.1

1.8
1.9000000000000001
2.0
2.1
```

The numbers from 1.8 to 2.1 are generated in increments of 0.1. We have two versions:

- In the first version, the integers from 18 to 21 are generated and then divided by 10.
- The second version starts with the first value and increases by 0.1 within the loop. Note that you must first calculate how often the loop must be run through.

3.4.7 Exercises

u_range

Consider the output of the following program (u_range.py file). Then check your results via a call:

```python
print("Loop 1")
for i in 2, 3, 6.5, -7:
    print(i)

print("Loop 2")
for i in range(3,11,3):
    print(i)

print("Loop 3")
for i in range(-3,14,4):
    print(i)

print("Loop 4")
for i in range(3,-11,-3):
    print(i)
```

u_range_inch

Write a program that generates the following output (u_range_inch.py file).

```
15 inches = 38.1 cm
20 inches = 50.8 cm
25 inches = 63.5 cm
30 inches = 76.2 cm
35 inches = 88.9 cm
40 inches = 101.6 cm
```

This list is a regular list of inch values, and the centimeter values are determined through conversion by a factor of 2.54.

3.4.8 Our Game: Version with range()

In the mental calculation game, the loop for repeating the input is now created using range(). At the same time, you have a counter that contains the consecutive number of the attempt. This value is used to output the number of attempts at the end of the program.

```python
import random
random.seed()

a = random.randint(1,10)
b = random.randint(1,10)
c = a + b
print(f"The task: {a} + {b}")

for attempt in range(1,10):
    print("Please enter a suggested solution:")
    number = int(input())

    if number == c:
        print(number, "is correct")
        break
    else:
        print(number, "is wrong")

print("Result:", c)
print("Number of attempts:", attempt)
```

Listing 3.28 game_range.py File

The output is:

```
The task: 5 + 1
Please enter a suggested solution:
7
```

```
7 is wrong
Please enter a suggested solution:
6
6 is correct
Result: 6
Number of attempts: 2
```

After calling the program, you have a maximum of nine attempts to enter the correct solution. The attempt variable serves as a counter for the attempts. After the correct solution has been entered (or after the loop has been completed), the number of attempts is displayed.

3.4.9   Loop with while

The while loop is used to control a repetition using a condition. In the following program, random numbers are added up and output. As long as the total of the numbers is less than 30, the process is repeated. If the total is equal to or greater than 30, the program will terminate.

```python
import random
random.seed()

subtotal = 0
while subtotal < 30:
    rnumber = random.randint(1,8)
    subtotal = subtotal + rnumber
    print(f"Number: {rnumber}, Subtotal: {subtotal}")
```

Listing 3.29 loop_while.py File

A possible output of the program would read as follows:

```
Number: 1, Subtotal: 1
Number: 8, Subtotal: 9
Number: 7, Subtotal: 16
Number: 4, Subtotal: 20
Number: 3, Subtotal: 23
Number: 8, Subtotal: 31
```

First, the variable for the total of the numbers is set to 0.

The while statement initiates the loop. The literal translation of the line is "as long as the total is less than 30," which refers to the following indented statements.

The word while is followed by a condition that is created using comparison operators. In this case, too, don't forget the colon at the end of the line, similar to if-else and for.

A random number is determined and added to the preceding total. The new total is therefore calculated from the old total plus the new random number. Then, the new total is output.

The end of the loop (and, in our case, the end of the program) is not reached until after the total has reached or exceeded the value 30.

3.4.10 Our Game: Version with while Loop and Counter

The while loop is now also used in the mental calculation game to repeat the input. After calling the program, you have any number of attempts to solve the task. The attempt variable, which serves as a counter for the attempts, must be controlled separately. It no longer automatically corresponds to the value of a loop variable.

```python
import random
random.seed()

a = random.randint(1,10)
b = random.randint(1,10)
c = a + b
print(f"The task: {a} + {b}")

number = c + 1
attempt = 0
while number != c:
    attempt = attempt + 1
    print("Please enter a suggested solution:")
    number = int(input())
    if number == c:
        print(number, "is correct")
    else:
        print(number, "is wrong")

print("Result:", c)
print("Number of attempts:", attempt)
```

Listing 3.30 game_while.py File

The output has not changed since the last version.

The while loop runs as long as the correct solution is not entered. The number variable is preset with a value that ensures that the while loop runs at least once. The attempt variable is preassigned with 0 and serves as a consecutive number. Once the correct solution has been entered, the number of attempts is displayed.

3.4.11 Exercise

u_while

Write a program (*u_while.py* file) that repeatedly prompts you to enter a value in inches after being called. The entered value should then be converted into centimeters and output. The program should terminate after the value 0 has been entered.

In a while loop, the condition according to which the loop is to be repeated is specified, not the condition according to which it is to be terminated. For this reason, you must formulate a condition in this program, namely, "as long as the input is not equal to 0."

3.4.12 Combined Assignment Expressions

Python 3.8 introduced the := operator for combined assignment expressions, also called the *walrus operator* due to its resemblance to the eyes and tusks of a walrus. This operator enables shorter expressions, which are, however, more complex.

Let's look at two loops, each of which outputs the values from 10 to 12:

```
a = 9
while a < 12:
    a = a + 1
    print(a)
print()

a = 9
while (a := a + 1) < 13:
    print(a)
```

Listing 3.31 loop_assignment.py File

In the header of the second loop, a is first increased by 1. Then, a is compared with 13. This comparison controls the loop. Due to the low precedence of the := operator, the assignment must be enclosed in parentheses. The output of the program reads:

```
10
11
12

10
11
12
```

3.5   Developing a Program

When developing your own programs, you should proceed step by step. First, give some thought to how the entire program should be structured, and you should do this thinking on paper. Which parts should go where in the sequence of steps? Do not try to write the entire program with all its complex components at once! This viewpoint is the biggest mistake that beginners (and sometimes advanced users) can make.

First, write a simple version of the first part of the program. Then, test it. Do not add the subsequent program section until you've performed a successful test. Test again after each change. If an error occurs, you'll know that the problem was caused by the last change. After the last addition, you have created a simple version of your entire program.

Now, change part of your program into a more complex version of itself. In this way, you can make your program more complex step by step until you've finally created the entire program as it corresponds to the initial considerations you put down on paper.

Sometimes, one or two changes to your design arise during the actual programming process. These changes are not a problem as long as the entire structure does not change. However, if changes accumulate, return to your paper mapping briefly and reconsider the structure. You don't always need to delete your earlier lines of code, maybe just change them a little and arrange them differently.

Write your programs clearly. If you're thinking about how to do three or four specific steps of your program at once, a better approach is to turn them into individual instructions that are executed in sequence. This method simplifies any troubleshooting. If you (or another person) change or expand your program later, building the program will be much faster.

You can use the print() function to check values and search for logical errors. In addition, you can temporarily mark individual lines of your program as comments to determine which part of the program is running without errors and which part is therefore prone to errors.

3.6   Errors and Exceptions

During the runtime of a program, errors may occur that you cannot foresee. If, for example, no valid number is entered after a prompt for a number, an exception occurs, and the program is terminated with an error message when the input is converted using int() or float().

Up to this point, we have assumed, for simplicity's sake, that correct entries have been made. In this section, I want to describe how you can avoid or intercept the consequences of incorrect entries.

Python versions 3.10 to 3.12 include many improvements to error messages. Now, error messages contain more information and are more accurate and therefore provide better help during troubleshooting. Suitable suggestions for rectifying an error are often already made.

3.6.1 Base Program

Let's look at a program designed to demonstrate an error.

```
print("Please enter an integer")
z = input()
number = int(z)
print(f"You have entered the integer {number}")
```

Listing 3.32 error_base.py File

If an incorrect entry is made (such as "3a"), the program aborts the conversion and generates the following output:

```
Please enter an integer
3a
Traceback (most recent call last):
  File "C:\Python\Examples\error_base.py", line 3, in <module>
    number = int(z)
ValueError: invalid literal for int() with base 10: '3a'
```

This information indicates the point in the program at which an error is detected (*error_base.py* file, line 3). The type of error is also communicated (ValueError).

3.6.2 Intercepting Errors

The first step is to intercept the consequences of an incorrect entry in order to prevent the program from aborting. For this task, you must determine the point at which an error could occur that Python can recognize. For this purpose, let's improve our program by making the following change:

```
print("Please enter an integer")
z = input()

try:
    number = int(z)
    print(f"You have entered the integer {number}")
except:
    print("Error converting the input")
```

Listing 3.33 error_intercept.py File

If a correct integer is entered, the program runs as before. If an incorrect entry is made, the program does not abort the conversion but instead displays the following message:

```
Please enter an integer
3a
Error converting the input
```

The try statement initiates exception handling. There are various branches that the program can run through (refer to Section 3.3 for more information on branches). The program tries to execute the statements that are indented after try. If the conversion is successful, the except branch will not be run, which is similar to the else branch of an if statement.

However, if the conversion is not successful, the error or *exception* can be intercepted via the except statement. In this case, all indented statements are executed in the except branch. The program runs to the end without aborting, as the error occurs but is intercepted.

You must place a colon after try and except.

> **Note**
>
> As only the *critical area* of your program will be included in the exception handling process, you should think about which parts of your program are prone to errors. The conversion of an input is one such critical point. Other possible errors are, for example, the editing of a file (which may not exist) or the output to a printer (which may not be switched on).

3.6.3 Outputting an Error Message

Different errors can occur in a larger critical area that is included in exception handling. It then makes sense to display the error message from the system. In the following program, this is shown by means of a small example:

```python
print("Please enter an integer")
z = input()

try:
    number = int(z)
    print(f"You have entered the integer {number}")
except Exception as e:
    print(e)
```

Listing 3.34 error_message.py File

When an error is intercepted, a reference to an error object is transferred to the e variable using the as keyword. The object is of the general error class Exception. Its value can be output. After entering "3a," you'll see the following output:

```
Please enter an integer
3a
invalid literal for int() with base 10: '3a'
```

3.6.4   Repeating an Entry

In our second step, we'll ensure that a new entry can be made after an incorrect entry. The entire input process with exception handling is repeated until the input is successful. Let's consider this improved example:

```
while True:
    print("Please enter an integer")
    z = input()
    try:
        number = int(z)
        print(f"You have entered the integer {number}")
        break
    except:
        print("Error converting the input")
```

Listing 3.35 error_input_new.py File

The following output is displayed when first an error is input and then when a correct value is input:

```
Please enter an integer
3a
Error converting the input
Please enter an integer
12
You have entered the integer 12
```

The truth value True is used to formulate an endlessly repeated while loop in which the input process with the exception handling is included. If the input is successful, the loop will be exited using break. If the entry is not successful, the entry process will be repeated.

3.6.5 Exercise

u_error

Improve our program for entering and converting any inch value into centimeters. The consequences of a possible input error should be intercepted. The program should repeat the prompt until the input is successful (*u_error.py* file).

3.6.6 Our Game: Version with Exception Handling

Let's add exception handling and use the `continue` keyword in our mental calculation game. Now, the consequences of an input error can be intercepted, and the program can continue as normal.

```python
import random
random.seed()

a = random.randint(1,10)
b = random.randint(1,10)
c = a + b
print(f"The task: {a} + {b}")

number = c + 1
attempt = 0
while number != c:
    attempt = attempt + 1
    print("Please enter an integer as a suggested solution:")
    z = input()

    try:
        number = int(z)
    except:
        print("You have not entered an integer")
        continue

    if number == c:
        print(number, "is correct")
    else:
        print(number, "is wrong")

print("Result:", c)
print("Number of attempts:", attempt)
```

Listing 3.36 game_exception.py File

The following output is generated:

```
The task: 7 + 6
Please enter an integer as a suggested solution:
12
12 is wrong
Please enter an integer as a suggested solution:
13a
You have not entered an integer
Please enter an integer as a suggested solution:
13
13 is correct
Result: 13
Number of attempts: 3
```

The conversion of the input is located in a try-except block. If the conversion is not successful due to an incorrect entry, the corresponding message appears. The rest of the loop is then skipped, and the next input is requested immediately.

3.7   Functions and Modules

Modularization, which is the subdivision of a program into functions, provides obvious advantages, especially for larger programs:

- Program parts that are required more than once are only defined once.
- Program parts can be used in several programs.
- Extensive programs are broken down into clear parts.
- Maintenance and servicing of programs is made easier.
- The program code is easier to understand for everyone involved, both during initial development and during subsequent improvements.

Python has numerous predefined functions that take a lot of the work out of development time. They are either built-in functions or functions made available via the integration of special modules.

For example, you're already using the built-in input() function. Each function has a specific task. The input() function stops the program and accepts an input.

Many functions have a what is called a *return value*. They return a result to the point in the program from which they are called. In the case of input(), the return value of this function is the character string entered.

> **Note**
>
> If you have already worked with another programming language, keep in mind that functions cannot be *overloaded* in Python. If you define a function multiple times, possibly with different parameters, only the last definition applies.

3.7.1 Simple Functions

Simple functions always perform the same action when they are called. In the following example, each call of the asterisk() function results in a visual separation being displayed on the screen:

```
# Definition of the function
def asterisk():
    print("-------------------")
    print("*** Separation ****")
    print("-------------------")

# Program
x = 12
y = 5
asterisk()                      # 1st Call
print("x =", x, ", y =", y)
asterisk()                      # 2nd Call
print("x + y =", x + y)
asterisk()                      # 3rd Call
print("x - y =", x - y)
asterisk()                      # 4th Call
```

Listing 3.37 function_simple.py File

The output of our function is shown in Figure 3.1.

First, we'll define the asterisk() function. The def keyword is followed by the name of the function, then parentheses and the usual colon. Values could be passed to the function within the parentheses, which we'll look more closely later in this chapter, from Section 3.7.2 onwards. The indented statements that follow are executed each time the function is called.

Initially, a function is only defined and made available for later use. Our sample program contains some calculations and outputs. The four calls of the asterisk() function are used to visually separate the outputs. After processing the function, the program continues with the statement that follows the function call.

You can call a function by writing its name followed by parentheses. If you want to pass information to the function, you must include this information (through what's called *parameters*) inside the parentheses.

```
--------------------
*** Separation ****
--------------------
x = 12 , y = 5
--------------------
*** Separation ****
--------------------
x + y = 17.
--------------------
*** Separation ****
--------------------
x - y = 7
--------------------
*** Separation ****
--------------------
```

Figure 3.1 Simple Functions

In general, you're largely free to choose the name of a function. The same rules apply as for the names of variables (see also Chapter 2, Section 2.1.6): In summary, a name can consist of the letters "a" to "z", "A" to "Z", numbers, and the _ character (underscore). A name must not begin with a digit and must not correspond to a reserved word in Python.

3.7.2 Functions with One Parameter

When you call a function, you can also transmit information through *parameters*. These parameters are analyzed within the function and lead to different results with different values. Let's consider an example.

```
# Definition of the function
def square(x):
    q = x*x
    print(f"Number: {x}, Square: {q}")

# Program
square(4.5)
a = 3
square(a)
square(2*a)
```

Listing 3.38 parameter.py File

The following output is generated:

```
Number: 4.5, Square: 20.25
Number: 3, Square: 9
Number: 6, Square: 36
```

The definition of the `square()` function contains a variable within the parentheses. When called, a value is transferred to the function and assigned to this variable. In our example, the following values are used:

- The number 4.5: The x variable is assigned the value 4.5 in the function.
- The value of the a variable: The x variable receives the current value of a in the function (i.e., 3).
- The result of a calculation: The x variable receives the current value of 2 * a in the function (i.e., 6).

> **Note**
>
> This function expects exactly one value. You cannot call this function without a value or with more than one value; otherwise, the program will abort with an error message.

3.7.3 Exercise

> **u_parameter**
>
> Let's say we need to calculate the taxes for different salaries (*u_parameter.py* file). If a salary is over $2,500, then 22% tax is payable; otherwise, 18%. This time, the calculation and output of the tax should take place within a function called `tax()`. The function should be called for the following salaries: $1,800, $2,200, $2,500, and $2,900.

3.7.4 Functions with Multiple Parameters

You can also transfer multiple parameters to a function. Make sure that the number of parameters matches and that they are in the correct order. Let's look at an example.

```
# Definition of the function
def calculation(x,y,z):
    result = (x+y) * z
    print("Result:", result)

# Program
calculation(2,3,5)
calculation(5,2,3)
```

Listing 3.39 parameters_multiple.py File

The following output is generated:

```
Result: 25
Result: 21
```

Exactly three parameters are expected, and three values are transmitted for both calls. From the result, notice how the order of the parameters is important:

- In the first call, x is assigned the value 2, y is assigned the value 3, and z is assigned the value 5. These assignments result in the following calculation: (2 + 3) * 5 = 25.
- The second call transfers the same numbers, but in a different order. The result is the following calculation: (5 + 2) * 3 = 21.

3.7.5 Functions with Return Value

Functions are often used to calculate and return results. In a program, this *return value* is either saved in a variable or output directly.

In Python, functions can return more than one return value (see Chapter 5, Section 5.6.4). In this section, however, we'll initially only look at functions that provide exactly one return value.

Let's define such a function and then call it several times.

```
# Definition of the function
def mean(x,y):
    result = (x+y) / 2
    return result

# Program
c = mean(3, 9)
print("Mean:", c)

x = 5
print("Mean:", mean(x,4))

y = -5.1
z = 2.8
print(f"Mean: {mean(y,z)}")
```

Listing 3.40 return_value.py File

The following output is generated:

```
Mean: 6.0
Mean: 4.5
Mean: -1.15
```

The result is first calculated within the function. This result is then returned to the caller using the return statement. The return statement then immediately ends the sequence of the function.

When the mean() function is called for the first time, the return value is stored temporarily in the c variable. This variable can be used at any point in the further course of the program.

Several actions are carried out on the second call: The mean() function is called and returns a result. The result is output immediately.

The third call shows that the call of a function can be embedded in a character string using a string literal. Then, the value of the expression in the curly brackets is output (i.e., the return value of the function).

3.7.6 Our Game: Version with Functions

The next version of our mental calculation game includes two functions, which are used to determine the task and to analyze the input.

```
# task() function
def task():
    a = random.randint(1,10)
    b = random.randint(1,10)
    print(f"The task: {a} + {b}")
    return a + b

# comment() function
def comment(inputnumber, result):
    if inputnumber == result:
        print(inputnumber, "is correct")
    else:
        print(inputnumber, "is wrong")

# Program
import random
random.seed()
c = task()
number = c + 1
attempt = 0

while number != c:
    attempt = attempt + 1
    print("Please enter a suggested solution:")
    z = input()
```

```
try:
    number = int(z)
except:
    print("You have not entered an integer")
    continue
comment(number,c)

print("Result:", c)
print("Number of attempts:", attempt)
```

Listing 3.41 game_function.py File

The two random numbers are determined in the task() function, and the task is displayed on the screen. You can also use return to return the result of a calculated expression (in this case, the result of the task) as a return value.

The comment() function is sent two numbers as parameters: the suggested solution entered and the correct result. The entered solution is examined within the function, and a corresponding comment is output. The function has no return value.

3.7.7 Exercises

u_return_value

Let's rewrite the program from the u_parameter exercise (Section 3.7.3). The tax is to be calculated within the tax() function and returned to the main program. The value should be output in the main program (*u_return_value.py* file).

u_conditional

Rewrite the program from exercise u_return_value. The branch in the function is created using a conditional expression. Its result is returned immediately using return. The function now only contains one statement (*u_conditional.py* file).

3.8 Type Hints

Since Python 3.5, you can use *type hints* to specify the data type of the parameters and the return value of a function.

Since Python 3.6, you can generally assign a data type to variables through type hints. The rules for type hints have been refined with every Python version since their introduction. Python 3.12 includes a number of enhancements as well.

Type hints can improve the readability of your code. Some programs and development environments can perform type checks. During these checks, incorrect type assignments lead to error messages. These messages make it easier to find errors during the development of a program.

> **Note**
>
> Type hints are controversial in parts of the Python developer community. An important basic rule still applies to Python: "Python will remain a dynamically typed language."

In the following, I assume that you use one of the two type checking programs, *pyright* (on Windows) or *mypy* (on Ubuntu Linux and macOS). These programs can be installed in the following ways (see also Appendix A, Section A.1):

- On Windows: `pip install pyright`
- On Ubuntu Linux: `sudo apt install mypy`
- On macOS: `pip install mypy`

3.8.1   Variables and Type Checks

Type hints are used in the next program.

```
a:int = 42
print("Value:", a, "Type:", type(a))

b:float = 42.5
print("Value:", b, "Type:", type(b))

c:str = "Hello"
print("Value:", c, "Type:", type(c))

d:bool = True
print("Value:", d, "Type:", type(d))

a = 4.5
b = 4
c = False
d = "abc"
```

Listing 3.42 type_hints_variable.py File

The name of a variable is followed by its desired type and a colon. The four variables of different types are output together with their respective type.

The four assignments at the end of the program change the type of the various variables. Thanks to Python's dynamic typing, this change does not trigger any error.

The output of the program reads as follows:

```
Value: 42 Type: <class 'int'>
Value: 42.5 Type: <class 'float'>
Value: Hello Type: <class 'str'>
Value: True Type: <class 'bool'>
```

Let's now examine this program using one of the type checkers (*pyright* or *mypy*). In Chapter 2, Section 2.3.2 and Section 2.3.3, I described how you can get to the command line or terminal. As shown in Figure 3.2, run one of the following commands:

```
pyright type_hints_variable.py
```

or

```
mypy type_hints_variable.py
```

```
C:\Python\Examples>pyright type_hints_variable.py
c:\Python\Examples\type_hints_variable.py
  c:\Python\Examples\type_hints_variable.py:13:5 - error: Expression of type
"float" is incompatible with declared type "int"
    "float" is incompatible with "int" (reportAssignmentType)
  c:\Python\Examples\type_hints_variable.py:15:5 - error: Expression of type
"Literal[False]" is incompatible with declared type "str"
    "Literal[False]" is incompatible with "str" (reportAssignmentType)
  c:\Python\Examples\type_hints_variable.py:16:5 - error: Expression of type
"Literal['abc']" is incompatible with declared type "bool"
    "Literal['abc']" is incompatible with "bool" (reportAssignmentType)
3 errors, 0 warnings, 0 informations
```

Figure 3.2 Wrong Type on Assignment

Three errors are reported due to incorrect type assignments. Only the b = 4 assignment does not lead to an error, as the integer value 4 is automatically converted into a float value.

3.8.2   Functions and Type Checks

There are generic functions in many programming languages (including Python). This kind of function can be executed for objects of different data types but produces a similar result.

In many other programming languages, a generic function requires a special form of definition. In the dynamically typed Python language, every function is initially generic, as the parameters only receive their type when the function is called. The rest of the function should match the various possible data types.

The types of the parameters and the return value of a function can also be checked using type hints.

Let's look at our first version of this sample program:

```
def add(x, y):
    z = x + y
    return z


print("int:", add(12, 5))
print("float:", add(5.2, 7.1))
print("str:", add("Man", "hattan"))
```

Listing 3.43 type_hints_function.py File, Version 1

The add() function is a generic function. It can perform its task for different types of parameters and return values.

First, the total of two int values or two float values is determined and returned. Then, the result of a concatenation of two str values is returned.

The output reads as follows:

```
int: 17
float: 12.3
str: Manhattan
```

In a second version of the program, type hints are used for parameters and the return value:

```
def add(x:float, y:float) -> float:
    z = x + y
    return z


print("int:", add(12, 5))
print("float:", add(5.2, 7.1))
print("str:", add("Man", "hattan"))
```

Listing 3.44 type_hints_function.py File, Version 2

The desired type of a parameter follows its name and a colon. The desired type of return value follows after the -> operator. In our example, the two parameters and the return value are supposed to have the float type.

Only the function header has been changed. Calling the program leads to the same result as before. However, the type check for this program results in the error message shown in Figure 3.3; the function may not be called with two str values.

```
C:\Python\Examples>pyright type_hints_function.py
c:\Python\Examples\type_hints_function.py
   c:\Python\Examples\type_hints_function.py:8:19 - error: Argument of type "Literal
['Man']" cannot be assigned to parameter "x" of type "float" in function "add"
      "Literal['Man']" is incompatible with "float" (reportArgumentType)
   c:\Python\Examples\type_hints_function.py:8:26 - error: Argument of type "Literal
['hattan']" cannot be assigned to parameter "y" of type "float" in function "add"
      "Literal['hattan']" is incompatible with "float" (reportArgumentType)
2 errors, 0 warnings, 0 informations
```

Figure 3.3 Incorrect Parameter Type

3.9   The Finished Game

At the end of this programming course, we want to extend our mental calculation game a little further by using many of the programming tools that are now available to you. Let's focus on these enhancements:

- Up to ten tasks are set one after the other. You can determine the number yourself while using the program.
- Besides addition, the other basic arithmetic operations are also used, namely, subtraction, multiplication, and division.
- The ranges from which the random numbers are selected depend on the calculation type. Multiplication, for example, uses smaller numbers than addition.
- A maximum of three attempts per task is allowed.
- The number of correctly solved tasks is determined.

The individual sections of our enhanced program are numbered. These numbers can be found in the description. Further additions will be made in later chapters. As a result, we get the following program:

```
# 1: Random generator
import random
random.seed()

# 2: Number of tasks
tasknumber = -1
while tasknumber<1 or tasknumber>10:
    try:
        print("How many tasks (1 to 10):")
        tasknumber = int(input())
    except:
        continue

# 3: Number of correct results
correct = 0
```

```
# 4: Loop with desired number of tasks
for task in range(1,tasknumber+1):

    # 5: Operator selection
    opnumber = random.randint(1,4)

    # 6: Operand selection
    if opnumber == 1:
        a = random.randint(-10,30)
        b = random.randint(-10,30)
        op = "+"
        c = a + b
    elif opnumber == 2:
        a = random.randint(1,30)
        b = random.randint(1,30)
        op = "-"
        c = a - b
    elif opnumber == 3:
        a = random.randint(1,10)
        b = random.randint(1,10)
        op = "*"
        c = a * b

    # 7: Division as a special case
    elif opnumber == 4:
        c = random.randint(1,10)
        b = random.randint(1,10)
        op = "/"
        a = c * b

    # 8: Task definition
    print(f"Task {task} of {tasknumber}: {a} {op} {b}")

    # 9: Loop with 3 attempts
    for attempt in range(1,4):
        # 10: Input
        try:
            print("Please enter a suggested solution:")
            number = int(input())
        except:
            # Conversion was not successful
            print("You have not entered an integer")
            # Continue loop immediately
            continue
```

```
    # 11: Comment
    if number == c:
        print(number, "is correct")
        correct = correct + 1
        break
    else:
        print(number, "is wrong")


    # 12: Correct result of the task
    print("Result:", c)


# 13: Number of correct results
print(f"Correct: {correct} of {tasknumber}")
```

Listing 3.45 game_finished.py File

The following output is generated:

```
How many tasks (1 to 10):
2
Task 1 of 2: 7 - 16
Please enter a suggested solution:
-9
-9 is correct
Result: -9
Task 2 of 2: 15 - 26
Please enter a suggested solution:
-10
-10 is wrong
Please enter a suggested solution:
-9
-9 is wrong
Please enter a suggested solution:
-12
-12 is wrong
Result: -11
Correct: 1 of 2
```

After initializing the random generator (1), the desired number of tasks is loaded (2). Because the player could have input an error, exceptions are handled. The counter for the number of correctly solved tasks (correct) is set to 0 (3). An outer for loop is started with the desired number (4).

The operator is determined by a random generator (5). There are different ranges for each operator from which the numbers are selected (6). The operator itself and the result are saved.

A special feature should be noted when it comes to division operations (7): You should only use integers. The two random operands (a and b) are therefore determined from the result of multiplication.

Then, the task is set (8). The sequential number and the total number of tasks are also displayed for better orientation. An inner for loop is started for a maximum of three attempts (9).

The entries made during use (10) are commented on (11). The correct result is displayed after a maximum of three attempts (12). Finally, the number of correctly solved tasks is displayed (13).

Chapter 4
Data Types

In Python, all data is stored in objects. This chapter deals with the properties and advantages of the different data types that characterize each object. We'll introduce you to operations, functions, and operators for each data type. A separate section on object references and object identity will complete our discussion of data types and objects. First, we'll focus on numbers, followed by strings (also known as *character strings*), lists, tuples, dictionaries, and sets. The similarities and differences among the data types are explained in detail.

4.1 Numbers

This section deals with integers, numbers with decimal places, fractions, and operations with numbers. Some built-in functions exist for numbers. The math module contains a range of mathematical functions for performing calculations.

4.1.1 Integers

The data type for integers is int. Numbers of this type are used with mathematical precision.

Usually, the decimal number system with the base 10 is used. The following number systems are also available in Python:

- The dual number system (with base 2)
- The octal number system (with base 8)
- The hexadecimal number system (with base 16)

Let's look at some examples.

```
a = 27
print("Decimal:", a)
print("Hexadecimal:", hex(a))
print("Octal:", oct(a))
print("Dual:", bin(a))
```

```
b = 0x1a + 12 + 0b101 + 0o67
print("Sum:", b)
```

Listing 4.1 number_integer.py File

The following output is generated:

```
Decimal: 27
Hexadecimal: 0x1b
Octal: 0o33
Dual: 0b11011
Sum: 98
```

The decimal number 27 is converted into the three other number systems, and then these results are output.

The hex() function is used to convert the number into the hexadecimal system. In addition to the digits 0 to 9, this system uses the letters "a" to "f" (or "A" to "F") as digits for the values from 10 to 15. The number 0x1b corresponds to the following value:

$1 \times 16^1 + B \times 16^0 = 1 \times 16^1 + 11 \times 16^0 = 16 + 11 = 27$

The oct() function is used to convert the number to the octal system. The octal system only uses the digits 0 to 7. For example, the number 0o33 corresponds to the following value:

$3 \times 8^1 + 3 \times 8^0 = 24 + 3 = 27$

The bin() function is used to convert the number into the dual (or binary) system. This system only uses the digits 0 and 1. The number 011011 corresponds to the following value:

$1 \times 2^4 + 1 \times 2^3 + 0 \times 2^2 + 1 \times 2^1 + 1 \times 2^0 = 16 + 8 + 2 + 1 = 27$

You can also calculate directly with numbers in other number systems. The calculation of the b variable results in:

0x1a + 12 + 0b101 + 0o67 =

$(1 \times 16^1 + A \times 16^0) + (12) + (1 \times 2^2 + 0 \times 2^1 + 1 \times 2^0) + (6 \times 8^1 + 7 \times 8^0) =$

$(16 + 10) + (12) + (4 + 1) + (48 + 7) = 98$

For the entry or assignment, the prefix 0x, 0b, or 0o must precede the other digits so that the corresponding number system is recognized.

At the lowest level, numbers are made up of bits and bytes. In Section 4.1.9, you'll work a little more intensively with dual numbers; the bin() function; and *bit operators*, which make it easier for you to access the bit level.

4.1.2 Numbers with Decimal Places

The data type for numbers with decimal places is called float. Called *floating point numbers*, these numbers are specified using a decimal point and, if necessary, in exponential notation.

Let's consider a small example.

```
a = 7.5
b = 2e2
c = 3.5E3
d = 4.2e-3
e = 1_250_000.500_001

print(a, b, c, d, e)
```

Listing 4.2 number_decimal_place.py File

The output reads:

```
7.5 200.0 3500.0 0.0042 1250000.500001
```

The a variable is assigned the value 7.5. The decimal places follow the decimal point. This rule also applies to entering a number with decimal places using the input() function. The b variable is assigned the value 200 (= 2×10^2 = 2 × 100). The c variable is given the value 3,500 (= 3.5×10^3 = 3.5 × 1,000); the d variable, the value 0.0042 (= 4.2×10^{-3} = 4.2 × 0.001).

For assignments in exponential notation, the character "e" (or "E") is used to express by how many digits and in which direction the decimal point is shifted within the number. This notation is suitable for very large or very small numbers because it saves you from having to enter lots of zeros.

Since Python 3.6, you can use an underscore to make numbers with many digits more readable. Inserting this character makes sense after every third digit, as was done with the variable e.

4.1.3 Exponential Operator **

The exponential operator ** is used to calculate powers, that is, expressions in the form *base to the power of exponent*. Let's consider some examples.

```
z = 5 ** 3
print("5 to the power of 3 =", z)
z = -5.2 ** -3.8
print("-5.2 to the power of -3.8 =", z)
z = 9 ** 0.5
```

```
print("Square root of 9 = 9 to the power of 1/2 =", z)
z = 27 ** (1.0/3.0)
print("Cube root of 27 = 27 to the power of 1/3 =", z)
```

Listing 4.3 number_power.py File

The following output is generated:

```
5 to the power of 3 = 125
-5.2 to the power of -3.8 = -0.0019018983172844654
Square root of 9 = 9 to the power of 1/2 = 3.0
Cube root of 27 = 27 to the power of 1/3 = 3.0
```

Both the base and the exponent can be positive or negative integers or numbers with decimal places. The exponential operator can also be used to calculate square roots and cube roots, for example.

4.1.4 Rounding and Conversion

The built-in round() function is used to round a number to any number of digits before or after the decimal point. In contrast, the int() function truncates the decimal places of a number. Let's look at some examples.

```
x = 1.499999
print("x:", round(x))
x = 1.500000
print("x:", round(x))
print()

x = 12/7
print("x:", x)
print("Rounded to six decimal places after the separator:", round(x,6))
print("Rounded to zero decimal places:", round(x))
print("int(x):", int(x))
print()

x = 12e6/7
print("x:", x)
print("Rounded to three decimal places before the separator:", round(x,-3))
print("Rounded to zero decimal places:", round(x))
print("int(x):", int(x))
```

Listing 4.4 number_round.py File

In this case, we round down to 1.499999 and round up from 1.500000 onwards.

The result of the division 12 / 7 is converted in three ways:

- The result is rounded to six decimal places after the separator using the built-in round() function.

- The same function is used to round the result to the next highest or next lowest integer.

- The result is converted into an integer using the built-in int() function.

The same happens with the result of the division 12000000 / 7:

- Due to the use of round(), the result is rounded to three decimal places before the separator.

- The same function is used to round the result to the next highest or next lowest integer.

- Using int(), the result is converted into an integer.

The following output is generated:

```
x: 1
x: 2

x: 1.7142857142857142
Rounded to six decimal places after the separator: 1.714286
Rounded to zero decimal places: 2
int(x): 1

x: 1714285.7142857143
Rounded to three decimal places before the separator: 1714000.0
Rounded to zero decimal places: 1714286
int(x): 1714285
```

4.1.5   Trigonometric Functions

In the math module, you'll find the trigonometric functions sin(), cos(), and tan() as well as the inverse trigonometric functions asin(), acos(), and atan(), among other things.

Let's consider a sample program.

```
import math

x = 30
xbm = math.radians(x)
print(f"Sine {x} Degree: {math.sin(xbm)}")
print(f"Cosine {x} Degree: {math.cos(xbm)}")
print(f"Tangent {x} Degree: {math.tan(xbm)}")
```

```
z = 0.5
print(f"Arcsine {z} in Degrees: {math.degrees(math.asin(z))}")
z = 0.866
print(f"Arccosine {z} in Degrees: {math.degrees(math.acos(z))}")
z = 0.577
print(f"Arctangent {z} in Degrees: {math.degrees(math.atan(z))}")
```

Listing 4.5 number_angle.py File

The following output is generated:

```
Sine 30 Degree: 0.49999999999999994
Cosine 30 Degree: 0.8660254037844387
Tangent 30 Degree: 0.5773502691896257
Arcsine 0.5 in Degrees: 30.000000000000004
Arccosine 0.866 in Degrees: 30.002910931188026
Arctangent 0.577 in Degrees: 29.984946007397852
```

After importing the math module, the sine, cosine, and tangent of the 30-degree angle are calculated first. All functions expect angles in radians. For this reason, a conversion from degrees to radians takes place beforehand using the radians() function.

The arcsine, arccosine, and arctangent are then calculated for specific values. These functions also provide an angle in radians. This result is then converted to degrees using the degrees() function.

4.1.6   Other Mathematical Functions

The math module contains a range of functions and constants. Let's consider a sample program.

```
import math

print("Square root of 64:", math.sqrt(64))
print("Cube root of -27:", math.cbrt(-27))
print("Integer square root of 80:", math.isqrt(80))
print()

print("Natural logarithm of 33:", math.log(33))
print("e to the power of 3.5:", math.exp(3.5))
print("2 to the power of 10:", math.exp2(10))
print()

print("Logarithm base 10 of 0.001:", math.log10(0.001))
print("10 ** -3:", 10 ** -3)
```

```
print()

print("Mathematical constant pi:", math.pi)
print("Euler's number e:", math.e)
print()

t1 = 3, 2, 5
t2 = 2, 4, 3
print("Product:", math.prod(t1))
print("Sum of products:", math.sumprod(t1, t2))

print("Factorial of 5:", math.factorial(5))
print("Greatest common divisor of 60 and 135:", math.gcd(60, 135))
print("Remainder:", math.remainder(10.9, 2.5))
print("Remainder:", math.remainder(11.9, 2.5))
print()

if not math.isclose(3, 2.96, rel_tol=0.01):
    print("Not close")

if math.isclose(3, 2.97, rel_tol=0.01):
    print("Close")
```

Listing 4.6 number_calculator.py File

The following output is generated:

```
Square root of 64: 8.0
Cube root of -27: -3.0
Integer square root of 80: 8

Natural logarithm of 33: 3.4965075614664802
e to the power of 3.5: 33.11545195869231
2 to the power of 10: 1024.0

Logarithm base 10 of 0.001: -3.0
10 ** -3: 0.001

Mathematical constant pi: 3.141592653589793
Euler's number e: 2.718281828459045

Product: 30
Sum of products: 29
Factorial of 5: 120
Greatest common divisor of 60 and 135: 15
```

```
Remainder:  0.9000000000000004
Remainder:  -0.5999999999999996
```

```
Not close
Close
```

The mathematical square root of a positive number can be calculated using the sqrt() function. Since Python 3.11, the cube root (i.e., the third root) can be calculated from a number using the cbrt() function. Since Python 3.8, you can use the isqrt() function to calculate the integer square root of a number. This result is the largest integer value that is smaller than the mathematical square root of a number.

The log() function calculates the natural logarithm of a positive number for the base e. The exp() function determines the mathematical inverse: the value of e^x. Since Python 3.11, the exp2() function can be used to determine the value of 2^x.

The log10() function calculates the logarithm of a positive number for the base 10. After that, the mathematical inverse is calculated, namely, the value of 10^x.

You also have the mathematical constants pi and e.

Since Python 3.8, the prod() function can be used to determine the product of the elements of an iterable, in this case, a tuple. The math looks as follows: 3 x 2 x 5 = 30. Since Python 3.12, you can use the sumprod() function to calculate the sum of the products of the element pairs of two iterables in the same position. The math looks as follows: 3 x 2 + 2 x 4 + 5 x 3 = 29. You'll learn more about tuples in Section 4.4.

The value of the factorial can be mathematically calculated only from positive integers. The factorial() function is used for this purpose.

The gcd() function has been available since Python 3.5. This function determines the *greatest common divisor (GCD)* of two integers, which is the largest number by which the two numbers can be divided without a remainder.

Since Python 3.7, the remainder of a division can be calculated in accordance with the Institute of Electrical and Electronics Engineers (IEEE) 754 standard using the remainder() function. This value is the difference to the next integer. What does this mean? In this example, 10.9 or 11.9 is divided by 2.5. The mathematical result lies between the two integers 4 (4 x 2.5 = 10) and 5 (5 x 2.5 = 12.5). In the case of 10.9, the value 10 is closer, so the remainder() function returns the value 0.9 (= 10.9 - 10). In the case of 11.9, the value 12.5 is closer, so the remainder() function returns the value -0.6 (= 11.9 - 12.5).

Since Python 3.5, you can use the isclose() function to determine whether two numbers are close to each other. The proximity of two numbers is determined using a relative tolerance, in this case, 0.01. In other words, the two figures can differ by a maximum of 1%. You can use one of two named parameters. Instead of rel_tol, you

can use abs_tol for a measurement with an absolute tolerance. More on named parameters follows in Chapter 5, Section 5.6.2.

4.1.7 More Precise Figures with Decimal Places

In the remaining subsections of Section 4.1, I want to focus on special topics relating to
numbers: more precise numbers with decimal places, complex numbers, bit operators,
and fractions. You can skim over these topics for now and look them up later, if necessary.

The Decimal class from the decimal module allows you to perform calculations with
numbers that have a large number of decimal places. You must keep a few special features in mind.

Let's first look at an example program.

```
import decimal

a = 10 / 7
print("float value:   ", a)
a = decimal.Decimal(10) / 7
print("Decimal value:", a)
print("Type:", type(a))
print()

print("Integers:", a + 2 * 3 - (12 - 2) // 2)
print("Numbers with decimal places:", a + decimal.Decimal(2.5))
print("Division:", a + decimal.Decimal(4 / 2))
print()

a = decimal.Decimal(24)
print("Value of a:", a)
print("Square root of a:", a.sqrt())
print()

print("Natural logarithm of a:", a.ln())
print("e to the power of 3.178:", decimal.Decimal(3.178).exp())
print()

print("Logarithm base 10 of a:", a.log10())
print("Reverse:", 10 ** decimal.Decimal(a.log10()))
```

Listing 4.7 number_decimal.py File

First, the output of the program is as follows:

```
float value:    1.4285714285714286
Decimal value: 1.428571428571428571428571429
Type: <class 'decimal.Decimal'>

Integers: 2.428571428571428571428571429
Numbers with decimal places:  3.928571428571428571428571429
Division: 3.428571428571428571428571429

Value of a: 24
Square root of a: 4.898979485566356196394568149

Natural logarithm of a: 3.178053830347945619646941601
e to the power of 3.178: 23.99870810642115598337164669

Logarithm base 10 of a: 1.380211241711606022936244587
Reverse: 23.99999999999999999999999998
```

The decimal module is imported so that the Decimal class becomes available. Decimal() is the constructor of the Decimal class. This constructor offers various options for creating an object of the Decimal class and returns a reference to this object.

In our example, a Decimal object with the integer value 10 is created. This value is then divided by the integer value 7. The result is much more accurate than the float division of 10 / 7.

Decimal objects, integers, and parentheses can be processed together within an expression as long as the operators involved produce integer results. This rule applies with the operators for addition, subtraction, multiplication, and integer division.

If numbers with decimal places or results of divisions are involved, they must first be converted into Decimal objects.

The following methods, among others, can be called for Decimal objects:

- sqrt() for the square root
- ln() for the natural logarithm with base e
- exp() for the value of e^x
- log10() for the logarithm with base 10

The ** operator can also be used for exponentiation.

> **Note**
>
> A *method* is a function that can only be called for objects of a certain class. I will describe classes, instances, properties, methods, constructors, and other terms from the field of object-oriented programming (OOP) in more detail in Chapter 6.

4.1.8 Complex Numbers

In Python, you have the option of saving complex numbers and using them for calculations. However, foundational mathematics for understanding complex numbers is not the subject of this book. You can find information on this topic, for example, at *https://en.wikipedia.org/wiki/Complex_number*.

Let's consider a simple example:

```python
a = 2.5 - 4.2j
print(f"a = {a}, Type: {type(a)}")
print(f"Real part: {a.real}, Imaginary part: {a.imag}")
print("Absolute value:", abs(a))
print("Conjugated complex:", a.conjugate())
print()


b = 3.7j
print("b =", b)
print(f"Real part: {b.real}, Imaginary part: {b.imag}")
print("Absolute value:", abs(b))
print()
...
```

Listing 4.8 number_complex.py File, Part 1 of 2

The output of the first part of the program reads as follows:

```
a = (2.5-4.2j), Type: <class 'complex'>
Real part: 2.5, Imaginary part: -4.2
Absolute value: 4.887739763939975
Conjugated complex: (2.5+4.2j)

b = 3.7j
Real part: 0.0, Imaginary part: 3.7
Absolute value: 3.7
```

By assigning a complex number, a becomes an object of the complex class. A complex number is made up of a real and an imaginary part. The imaginary part is identified by the character "j" (or "J"). If only an imaginary part is assigned, the real part is set to 0. Complex numbers are displayed in parentheses. However, there's an exception: If the complex number only has an imaginary part, then it's displayed without parentheses (see variable b).

The real and imag properties provide the respective parts of the complex number. The abs() function returns the absolute value of the complex number. The conjugate() method returns the conjugated complex number (i.e., the complex number with a different sign for the imaginary part).

Objects (e.g., a); the properties of objects (e.g., a.real and a.imag); or function calls (e.g., type(a)) can be embedded in a string literal using curly brackets.

Let's now look at the second part of our program:

```
...
print("a + b =", a + b)
print("a - b =", a - b)
print("a * b =", a * b)
print("a / b =", a / b)
print("a ** 2.5 =", a ** 2.5)
print("5.1 + a / 3.2j * 2.8 =", 5.1 + a / 3.2j * 2.8)
print()

c = 2.5 - 4.2j
print("c =", c)
print("a == c:", a == c)
print("b != c:", b != c)
print()

c = 1j
print("c =", c)
print("c * c =", c * c)
```

Listing 4.9 number_complex.py File, Part 2 of 2

The output of the second part of the program reads as follows:

```
a + b = (2.5-0.5j)
a - b = (2.5-7.9j)
a * b = (15.540000000000001+9.25j)
a / b = (-1.135135135135135-0.67567567567756757j)
a ** 2.5 = (-44.83645966023058-27.915620445612213j)
5.1 + a / 3.2j * 2.8 = (1.4249999999999998-2.1875j)

c = (2.5-4.2j)
a == c: True
b != c: True

c = 1j
c * c = (-1+0j)
```

You can perform calculations with complex numbers using the operators +, -, *, /, and **, according to their corresponding mathematical rules. You can also calculate mixed expressions that contain real numbers as well as complex numbers. Only the == and

!= operators can be used to compare complex numbers. The square of the complex number j (i.e., 0 + 1j) corresponds to –1.

4.1.9 Bit Operators

All data, whether numbers or character strings, is made up of bits and bytes at the hardware level. At this level, you can work with dual numbers (see also Section 4.1.1), the bin() function, and what are called *bit operators*. Let's consider an example.

```python
# Only 1 bit set
bit0 = 1           # 0000 0001
bit3 = 8           # 0000 1000
print(bin(bit0), bin(bit3))

# Bitwise AND
a = 5              # 0000 0101
res = a & bit0     # 0000 0001
if res:
    print(a, "is odd")

# Bitwise OR
res = 0            # 0000 0000
res = res | bit0   # 0000 0001
res = res | bit3   # 0000 1001
print("Bits set in sequence:", res, bin(res))

# Bitwise exclusive OR
a = 21             # 0001 0101
b = 19             # 0001 0011
res = a ^ b        # 0000 0110
print("Unequal bits:", res, bin(res))

# Bitwise inversion, x becomes -(x+1)
a = 11             # 0000 1011
res = ~a           # 1111 0100
print("Bitwise inversion:", res, bin(res))

# Shift bitwise
a = 11             # 0000 1011
res = a >> 1       # 0000 0101
print("Shifted right by 1:", res, bin(res))
res = a << 2       # 0010 1100
print("Shifted left by 2:", res, bin(res))
```

Listing 4.10 operator_bit.py File

Here's the output:

```
0b1 0b1000
5 is odd
Bits set in sequence: 9 0b1001
Unequal bits: 6 0b110
Bitwise inversion: -12 -0b1100
Shifted right by 1: 5 0b101
Shifted left by 2: 44 0b101100
```

First, the `bit0` and `bit3` variables are introduced, which are required for some of the subsequent calculations. These variables have the values 1 and 8. At the end of the program line, you can see them as a dual number, that is, written using 8 bits (= 1 byte). The last bit of a byte is called bit 0, the penultimate bit is bit 1, and so on. The values of the `bit0` and `bit3` variables were selected in such a way that only one bit is set in each case (= 1), while the remaining bits are not set (= 0).

Imagine you have a row of eight LEDs that are either *on* or *off*. This information can be stored within one byte. If one of its bits is set, the relevant LED is *on*; otherwise, it is *off*. For clarification, the `bit0` and `bit3` variables are output as a dual number using the `bin()` function.

You can use the bit operator & to link two numbers bit by bit. Similar to the logical and operator (see Chapter 3, Section 3.3.8), a specific bit in the result will only be set if this bit is set in both numbers. This operation is carried out for each individual bit.

If you want to know whether a specific bit is set within a number, you can use the & bit operator to link this number with another number in which only this one bit is set. If it is bit 0, you also know whether the number is even (bit 0 = 0) or odd (bit 0 = 1).

The bit operator | is used to link two numbers bit by bit. Similar to the logical operator or (see also Chapter 3, Section 3.3.8), a specific bit is set in the result if this bit is set in one of the two numbers or in both numbers. This operation is also carried out for each individual bit.

If you want to set a single bit of a number, you can use the bit operator | to link this number with another number in which only this one bit is set.

The bit operator ^ is used for the bitwise connection of two numbers via an exclusive OR (XOR). A specific bit in the result is set if this bit is only set in one of the two numbers. If the bit is set in both numbers, the result bit is not set. This operation is also carried out for each individual bit.

You can use the bit operator ~ to invert a number bit by bit. The number x becomes the number -(x+1) (i.e., 11 becomes –12).

The two bit operators >> and << are used to shift bits within a number:

- You can shift all bits to the right by a certain number of places using >>. The bits that "fall out" to the right are lost. A shift by n bits to the right corresponds to an integer division by 2^n. A shift of 1 bit to the right therefore corresponds to an integer division by 2.

- You can use << to shift all bits a certain number of places to the left. A shift by n bits to the left corresponds to a multiplication by 2^n. A shift of 1 bit to the left therefore corresponds to a multiplication by 2.

4.1.10 Fractions

Python can perform calculations with fractions or provide information about fractions. For this purpose, the Fraction class from the fractions module is used.

Let's write a sample program in which some calculations with fractions are carried out according to the rules of mathematical fractions (see *https://en.wikipedia.org/wiki/Fraction*). At the end, this program will make some comparisons between fractions.

```
import fractions

b1 = fractions.Fraction(3, -2)
b2 = fractions.Fraction(1, 4)

b3 = b1 * b2
print(f"{b1} * {b2} = {b3}")
print(f"{b1} / {b2} = {b1 / b2}")
print(f"{b1} + {b2} = {b1 + b2}")
print(f"{b1} - {b2} = {b1 - b2}")
print(f"{b1} + {b2} * {b3}= {b1 + b2 * b3}")
print(f"({b1} + {b2}) * {b3}= {(b1 + b2) * b3}")

b4 = fractions.Fraction(-30, 20)
print(f"{b1} == {b4}: {b1 == b4}")
print(f"{b2} > {b4}: {b2 > b4}")
print(f"{b2} < {b4}: {b2 < b4}")
```

Listing 4.11 number_fraction_calculate.py File

A fraction is created using two integers as the numerator and denominator and is displayed with a forward slash as the fraction bar.

The Fraction() function from the fractions module provides various options for creating a fraction, or more precisely a Fraction object. Fraction() is the constructor of the Fraction class that creates an object of the class and returns a reference to this object.

Fraction objects are automatically reduced when they are created. If the fraction has exactly one negative sign in the numerator or denominator, it is placed before the numerator. If the fraction has two negative signs, the fraction is multiplied by –1, and the negative signs disappear.

Here's the output of the program:

```
-3/2 * 1/4 = -3/8
-3/2 / 1/4 = -6
-3/2 + 1/4 = -5/4
-3/2 - 1/4 = -7/4
-3/2 + 1/4 * -3/8= -51/32
(-3/2 + 1/4) * -3/8= 15/32
-3/2 == -3/2: True
1/4 > -3/2: True
1/4 < -3/2: False
```

Our next program illustrates some of the properties and methods of the `Fraction` object.

```
import fractions

b1 = fractions.Fraction(12, 28)
print("Fraction object:", b1, type(b1))
print(f"Numerator: {b1.numerator}, Denominator: {b1.denominator}")
print(f"Value: {b1:.12f}")
print()

x = 2.375
print("Number:", x)
b2 = fractions.Fraction(x)
print("Fraction object:", b2)
print()

b3 = fractions.Fraction("350_000 /280_000")
print("From string:", b3)
print("Integer:", b3.is_integer())
print(f"Formatted: {b3:.3f}")
```

Listing 4.12 number_fraction_object.py File

The output reads:

```
Fraction object: 3/7 <class 'fractions.Fraction'>
Numerator: 3, Denominator: 7
Value: 0.428571428571

Number: 2.375
Fraction object: 19/8
```

```
From string: 5/4
Integer: False
Formatted: 1.250
```

The numerator and denominator of the fraction are available individually in the numer-ator and denominator properties. Since Python 3.12, a fraction can be output using a for-matted string literal. In this case, the value of the fraction is displayed with 12 decimal places. You can find more information on formatting numbers in Chapter 5, Section 5.2.2.

You can create a fraction from a number with decimal places. This step turns the num-ber 2.375 into the Fraction object 19/8.

A fraction can be created from a string containing two integers separated by a forward slash. Since Python 3.12, this string can contain spaces. Since Python 3.11, the two num-bers can be made more readable using underscores (see also Section 4.1.2).

The is_integer() method has been available since Python 3.12. This method provides information as to whether the value of a fraction is an integer.

In our next program, a fraction is used to approximate a number with decimal places. The limit_denominator() method is used for this purpose.

```python
import fractions

x = 1.84953
print("Number:", x)

b3 = fractions.Fraction(x)
print("Fraction object:", b3)

b4 = b3.limit_denominator(100)
print("Approximated to denominator max. 100:", b4)

value = b4.numerator / b4.denominator
print("Value:", value)
print("rel. error:", abs((x-value)/x))
```

Listing 4.13 number_fraction_approximate.py File

The output is as follows:

```
Number: 1.84953
Fraction object: 8329542618810553/4503599627370496
Approximated to denominator max. 100: 172/93
Value: 1.8494623655913978
rel. error: 3.656843014286614e-05
```

The number 1.84953 is examined and corresponds to the fraction 184953/100000. The `limit_denominator()` method is used to limit the denominator to the number 100. Thus, you're looking for the fraction that has a denominator with a maximum value of 100 and comes closest to the number 1.84953.

In the present case, this fraction is 172/93. It has the value 1.8494623655913978 and is thus quite close to our original number. The relative error between this value and the analyzed number is only $3.65684301429 \times 10^{-5}$.

The relative error is determined using the built-in `abs()` function to calculate the absolute value of a number (i.e., the number without the sign).

If you're missing the `gcd()` method for determining the GCD, since Python 3.5, it has been part of the `math` module as a function (Section 4.1.6) and is designated as *deprecated* in the `fractions` module.

4.2   Character Strings

Character strings are sequences of individual characters. There are also other data types for which the rules of sequences apply. The following is an introduction to the rules of sequences using character strings.

4.2.1   Properties

Strings are objects of the `str` type. They are placed in single quotation marks ('), double quotation marks ("), and triple double quotation marks ("""). You can use a `for` loop to access the elements of a character string.

Let's consider an example:

```
ta = "The"
print("Text:", ta)
print("Type:", type(ta))
print("Number of characters:", len(ta))
for z in ta:
    print(z)
for i in range(len(ta)):
    print(f"{i}: {ta[i]}")

tb = 'This is also a character string'
print(tb)

tc = """This character string
        runs over
        several lines"""
```

```
print(tc)

td = 'There are "double quotation marks" saved here'
print(td)
```

Listing 4.14 text_property.py File

The following output is generated:

```
Text: The
Type: <class 'str'>
Number of characters: 3
T
h
e
0: T
1: h
2: e
This is also a character string
This character string
        runs over
        several lines
There are "double quotation marks" saved here
```

For the ta string, the type() function is used to output the data type, and the len() function is used to output the number of elements (i.e., the length of the string).

A string is an iterable that consists of individual elements. The elements are accessed using the first for loop.

The consecutive number of an element is designated by the *index*. You can access the element with the desired index using the square brackets [and].

This is done in the second for loop. It's run through using the loop variable i, which takes on the values from 0 to (length of character string) -1 one after the other. The tx[0], tx[1]... elements are therefore output one after the other, up to the tx[length-1] element.

The tb string is specified in single quotation marks. The tc string extends over several lines. For this purpose, it is noted in triple double quotation marks.

The td string illustrates the advantage of having several alternatives: The double quotation marks are part of the text and are also output.

4.2.2 Operators

The + operator is used to concatenate multiple sequences, while the * operator is used to multiply a sequence. You can use the in operator to determine whether a specific element is contained in a sequence. The not in operator states the opposite. Consider the following example of these operators, applied to strings:

```
circle = "-oooo-"
asterisk = "***"
line = asterisk + circle * 3 + asterisk
print(line)

tx = "Hello"
print("Text:", tx)
if "e" in tx:
    print("e is included")
if "z" not in tx:
    print("z is not included")
```

Listing 4.15 text_operator.py File

The output reads:

```
***-oooo--oooo--oooo-***
Text: Hello
e is included
z is not included
```

The line string is assembled using the concatenation operator + and the multiplication operator *. The -oooo- text is saved three times in a row in line.

4.2.3   Slices

Areas of sequences are referred to as *slices*. I want to illustrate the use of slices using an example string. Slices can also be applied to other sequences in the same way.

The Hello World string is saved as an example of a sequence. Table 4.1 displays the individual elements with the corresponding index. The index starts at 0 at the beginning of the sequence. Alternatively, you can also use a negative index, which starts with –1 at the end of the sequence.

Index	0	1	2	3	4	5	6	7	8	9	10
Element	H	e	l	l	o		W	o	r	l	d
Negative index	-11	-10	-9	-8	-7	-6	-5	-4	-3	-2	-1

Table 4.1 Sequence with Index

Like a single element, a slice can be specified using square brackets. A slice can start with a start index, followed by a colon and an end index. Let's look at a simple example.

```
tx = "Hello world"
print("Text:", tx)

print("[2:7]:", tx[2:7])
print("[:7]:", tx[:7])
print("[2:]:", tx[2:])
print("[:]:", tx[:])
print("[1]:", tx[1])
print("[1:-2]:", tx[1:-2])
print()

print("[2:7:2]:", tx[2:7:2])
s = slice(2, 7, 2)
print("[2:7:2]:", tx[s])
a = 2
b = 7
print("[2:7:2]:", tx[a:b:a])
```

Listing 4.16 text_slice.py File

The following output is generated:

```
Text: Hello world
[2:7]: llo w
[:7]: Hello w
[2:]: llo world
[:]: Hello world
[1]: e
[1:-2]: ello wor

[2:7:2]: low
[2:7:2]: low
[2:7:2]: low
```

Here's an explanation of the different slices:

- Slice [2:7]: The range extends from the element identified by the start index to directly before the element identified by the end index.
- Slice [:7]: If the start index is omitted, the range begins at 0.
- Slice [2:]: If the end index is omitted, the range ends at the end of the sequence.
- Slice [:]: If both indexes are omitted, the range extends over the entire sequence. This is required for the decomposition of multidimensional sequences (Section 4.3.2).
- Slice [1]: If only one index is specified, there is no slice, only a single element.

- Slice [1:-2]: If an index is specified with a negative number, this index is measured from the end, starting at –1.
- Slice [2:7:2]: After the start index and the end index, an integer increment can be specified with which the individual elements for the slice are compiled.
- The built-in slice() function returns a slice object. This object can be used to determine a slice of a sequence.
- The start index, end index, and increment can be specified using variables.

A slice of a sequence has the same data type as the sequence. In other words, a slice of a string is also a string; a slice of a list is also a list.

4.2.4 Mutability

Our next example shows that a string cannot be changed. No characters or slices can be replaced by other characters or slices.

```
tx = "This is a text"
print(tx)

try:
    tx[4:6] = "was"
except:
    print("Error")

tx = tx[:5] + "was" + tx[7:]
print(tx)
```

Listing 4.17 text_mutability.py File

The output reads:

```
This is a text
Error
This was a text
```

An attempt is made to replace part of the string with a slice, but this attempt ends in an error.

The only way to change a variable that contains a character string is to assign a new string to the same variable. The new string can be composed of parts of the old string.

4.2.5 Searching and Replacing

The next example illustrates some methods for searching and replacing data within texts.

```
tx = "This is a sample sentence"
print("Text:", tx)

find = "is"
print("Find text:", find)
print()

num = tx.count(find)
print(f"count: The string {find} occurs {num} times")

pos = tx.find(find)
while pos != -1:
    print("At position", pos)
    pos = tx.find(find, pos+1)

pos = tx.rfind(find)
if pos != -1:
    print("rfind: The last time at position", pos)
print()

if tx.startswith("This"):
    print("Text starts with This")
if not tx.endswith("This"):
    print("Text does not end with This")

tx = tx.replace("This", "That")
print("After replacement:", tx)

z = 48.2
tx = str(z)
tx = tx.replace(".", ",")
print("Number with comma:", tx)
```

Listing 4.18 text_search.py File

The following output is generated:

```
Text: This is a sample sentence
Find text: is

count: The string is occurs 2 times
At position 2
At position 5
rfind: The last time at position 5
```

```
Text starts with This
Text does not end with This
After replacement: That is a sample sentence
Number with comma: 48,2
```

The count() method returns the number of occurrences of a search text within the analyzed text.

The find() method returns the position at which a search text occurs in an analyzed text. If the search text does not occur, the value -1 will be returned.

You can optionally specify a second parameter in the find() method that determines the position from which the search is to start. You can use this parameter to find all occurrences of a search text using a while loop. The condition of the loop is "as long as the element appears in the (remaining) analyzed text." The first time this condition is used, the entire analyzed text is checked. Then, only the part of the analyzed text after the last position found will be checked.

The rfind() method returns the position of the last occurrence of the search text within the analyzed text.

You can use the startswith() and endswith() methods to check whether a character string begins or ends with a specific text. Both methods return a truth value, so the return value can be used to control the branch.

The replace() method replaces the partial text you are searching for with another partial text and returns the changed text for reassignment.

The built-in str() function creates a character string from an object. You can use this property, for example, to format a number individually by first converting the number into a string and then editing it with string functions.

4.2.6 Removing Blank Spaces

You may have blank spaces, tab characters (\t), or end-of-line characters (\n) at the beginning or end of a text, for example, after importing a text from a file. These characters can be removed using the strip() method. Let's look at some examples.

```
tx = "   \tHello world\n\t   "
print(f"|{tx}|")
print(f"|{tx.strip()}|")
```

Listing 4.19 text_blank.py File

To make things clearer, the | character is also displayed before and after the text. The output of the program is shown in Figure 4.1.

```
|        Hello world
          |
|Hello world|
```

Figure 4.1 Deleting Characters at the Beginning and End of the Text

4.2.7   Decomposing Texts

The split() method is used to decompose a text into a list of partial texts. In our next example, we'll look at two ways to use this method.

```
tx = "This is a sentence"
print("Text:", tx)
lt = tx.split()
for i in range(0, len(lt)):
    print("Element:", i, lt[i])
print()

tx = "Mayer;John;6714;3500.00;03/15/62"
print("Text:", tx)
lt = tx.split(";")
for i in range(0, len(lt)):
    print("Element:", i, lt[i])
```

Listing 4.20 text_decompose.py File

The output reads:

```
Text: This is a sentence
Element: 0 This
Element: 1 is
Element: 2 a
Element: 3 sentence

Text: Mayer;John;6714;3500.00;03/15/62
Element: 0 Mayer
Element: 1 John
Element: 2 6714
Element: 3 3500.00
Element: 4 03/15/62
```

In the split() method, by default, the space character is regarded as the separator.

The built-in len() function also returns the number of elements for a list. More information on lists can be found in Section 4.3.

The individual parts of a data record are often separated from each other by means of a semicolon. You can use this character as a parameter of the split() method to split a data record into its individual elements.

4.2.8   Constants

The string module provides some useful string constants, for example, all letters, all digits, or all punctuation marks, as illustrated in our next program.

```
import string
print("Lowercase:", string.ascii_lowercase)
print("Uppercase:", string.ascii_uppercase)
print("Letters:", string.ascii_letters)
print("Digits:", string.digits)
print("Punctuation marks:", string.punctuation)
```

Listing 4.21 text_constants.py File

These constants can be used, for example, to randomly select characters for a password (see Chapter 5, Section 5.4).

Properties and methods for character strings are available by default in Python. To use the constants of the string module, you must first import the module.

The output reads as follows:

```
Lowercase: abcdefghijklmnopqrstuvwxyz
Uppercase: ABCDEFGHIJKLMNOPQRSTUVWXYZ
Letters: abcdefghijklmnopqrstuvwxyzABCDEFGHIJKLMNOPQRSTUVWXYZ
Digits: 0123456789
Punctuation marks: !"#$%&'()*+,-./:;<=>?@[\]^_`{|}~
```

4.2.9   The bytes Data Type

The bytes data type is a special topic. You can skip this section for now and return later, if necessary.

The strings we've dealt with so far are objects of the str type. They are created from the extensive character set of Unicode characters.

Objects of the less frequently used bytes type are formed from a smaller character set. Their character code is only in the range between 0 and 255. Each character is only stored in a single byte. You can create bytes objects using byte literals or using the built-in bytes() function.

Byte literals start with a "b" or a "B." You must take this naming rule into account for creating entries or making assignments. The output is preceded by a b. Let's consider a sample program.

```
st = "Hello"
print(st, type(st))

by = b'Hello'
print(by, type(by))

by = bytes("Hello", "UTF-8")
print(by, type(by))

by = b'Hello'
st = by.decode()
print(st, type(st))
```

Listing 4.22 bytes.py File

This program generates the following output:

```
Hello <class 'str'>
b'Hello' <class 'bytes'>
b'Hello' <class 'bytes'>
Hello <class 'str'>
```

First, a str object and a bytes object are created via assignment and output along with their types. Note the b in front of the byte literal.

The built-in bytes() function then converts a str object into a bytes object. The encoding of the str object must be specified, in this case, UTF-8.

The decode() method is used to convert a bytes object into a str object.

4.3   Lists

In its most general form, a list is a sequence of objects. It can contain elements of different data types. A list offers many possibilities, including the functionality of one-dimensional and multidimensional fields (arrays), as they are known from other programming languages.

4.3.1   Properties and Operators

Unlike a string, a list can be changed. Apart from that distinction, a string is basically just a list for storing individual characters. Let's look at a few example lists.

```
import random

z = [3, 6.2, -12]
print("List:", z)
```

```
print("Element:", z[0])
print("Slice:", z[:2])
print()

print("Loop:")
for element in z:
      print(element)
print()

print("Loop with index:")
for i in range(len(z)):
      print(f"{i}: {z[i]}")
print()

a = ["Paris", "Lyon"]
b = ["Rome", "Pisa"]
c = a + b * 2
print("Added and multiplied:", c)

print("Random element:", random.choice(c))
random.shuffle(c)
print("After mixing:", c)
```

Listing 4.23 list_property.py File

A possible output could read as follows:

```
List: [3, 6.2, -12]
Element: 3
Slice: [3, 6.2]

Loop:
3
6.2
-12

Loop with index:
0: 3
1: 6.2
2: -12

Added and multiplied: ['Paris', 'Lyon', 'Rome', 'Pisa', 'Rome', 'Pisa']

Random element: Pisa
After mixing: ['Paris', 'Rome', 'Pisa', 'Pisa', 'Rome', 'Lyon']
```

A list is specified within square brackets. The individual elements are separated by commas.

The z variable contains a list of numbers. You can obtain a single element of a list using an index and a range using a slice.

A for loop can be used to output all elements of a list, with or without an index. You can determine the length of a sequence (i.e., the number of elements in a list) using the len() function.

Lists can be concatenated using the + operator and multiplied using the * operator.

The choice() function from the random module returns a random element from the list. The shuffle() function, also from the random module, mixes the elements of the list using a random generator.

4.3.2 Multidimensional Lists

A list can contain elements of different data types, while these elements can themselves be lists. You can use this characteristic to create a multidimensional list.

```
z = [["Paris","Fr",3.5], ["Rome","It",4.2], ["Madrid","Sp",3.2]]
print("List:", z)
print("Length:", len(z))
print()

print("Sublist:", z[0])
print("Length:", len(z[0]))
print("Slice of sublists:", z[:2])
print()

print("Element:", z[2][0])
print("Length:", len(z[2][0]))
print("Slice of elements:", z[2][:2])
print("Slice on the lowest level:", z[2][0][:3])
print()

for i in range(len(z)):
    print(f"{i}: {z[i][0]} has {z[i][2]} million inhabitants")
for city in z:
    print(f"{city[0]} has {city[2]} million inhabitants")
```

Listing 4.24 list_multidimensional.py File

The output reads:

```
List: [['Paris', 'Fr', 3.5], ['Rome', 'It', 4.2], ['Madrid', 'Sp', 3.2]]
Length: 3
```

```
Sublist: ['Paris', 'Fr', 3.5]
Length: 3
Slice of sublists: [['Paris', 'Fr', 3.5], ['Rome', 'It', 4.2]]

Element: Madrid
Length: 6
Slice of elements: ['Madrid', 'Sp']
Slice on the lowest level: Mad

0: Paris has 3.5 million inhabitants
1: Rome has 4.2 million inhabitants
2: Madrid has 3.2 million inhabitants
Paris has 3.5 million inhabitants
Rome has 4.2 million inhabitants
Madrid has 3.2 million inhabitants
```

This multidimensional z list has three elements, namely, three sublists. Each of these sublists consists of two character strings and a number. You can access both the entire list and the sublists using an index or a slice.

The len() function can return the length of a multidimensional list on several levels. If the function is called for the entire list, it returns the number of elements at the top level, in this case, the number of sublists. For a sublist, the function returns the number at the next level, in this case, the number of elements in the sublist.

You can access individual elements of a sublist by specifying several indexes. The first index represents the sublist, while the second index represents the element within the sublist.

You can use for loops to run through multidimensional lists in different ways.

4.3.3   Mutability

In contrast to strings, lists can be changed. You can delete individual elements or replace them with other elements or a list. You can also delete slices or replace them with a list, which can change the length or dimension of a list. Let's consider an example.

```
z = [12, 7, 38, -5]
print("List:", z)

z[2] = 16
print("Element replaced by element:", z)

z[1:3] = [43, -8, 72]
print("Slice replaced by list:", z)
```

```
del z[2]
print("Element removed:", z)

del z[2:]
print("Slice removed:", z)

z[0] = [-28, 19]
print("Element replaced by list:", z)
```

Listing 4.25 list_change.py File

In this example, the output describes the list in various states:

```
List: [12, 7, 38, -5]

Element replaced by element: [12, 7, 16, -5]
Slice replaced by list: [12, 43, -8, 72, -5]
Element removed: [12, 43, 72, -5]
Slice removed: [12, 43]
Element replaced by list: [[-28, 19], 43]
```

The list initially consists of four numbers and is changed in the following ways:

- A single element is replaced by a single element. The length of the list remains the same.
- A slice is replaced by a list. This inner list can have the same or a different length than the slice. The length of the outer list z may change.
- An element or a slice is removed from the list using the del keyword. The list gets shorter each time.
- A single element is replaced by a list, which creates a new sublist. The list gains an additional dimension.

A slice can only be replaced by an iterable. You therefore cannot replace a slice with a single element.

4.3.4   Methods

Our next program illustrates a number of other methods for changing and examining lists.

```
z = ["Paris", "Nantes", "Lyon", "Metz"]
print("List:", z)

z.sort()
print("Sort by text:", z)
```

```
print()

z = [12, 7, 38, -5]
print("List:", z)

z.sort()
print("Sort by numbers:", z)

z.insert(1, 16)
print("Insert:", z)

z.reverse()
print("Reverse:", z)

find = 16
if find in z:
    z.remove(find)
    print("Remove:", z)

z.append(12)
print("Add at the end:", z)
print()

find = 12
print(f"Number found:", z.count(find))

startpos = 0
while find in z[startpos:]:
    pos = z.index(12, startpos)
    print("Position:", pos)
    startpos = pos + 1
```

Listing 4.26 list_methods.py File

In this example, the output describes the list in various states:

```
List: ['Paris', 'Nantes', 'Lyon', 'Metz']
Sort by text: ['Lyon', 'Metz', 'Nantes', 'Paris']

List: [12, 7, 38, -5]

Sort by numbers: [-5, 7, 12, 38]
Insert: [-5, 16, 7, 12, 38]
Reverse: [38, 12, 7, 16, -5]
Remove: [38, 12, 7, -5]
```

```
Add at the end: [38, 12, 7, -5, 12]

Number found: 2
Position: 1
Position: 4
```

First, a list of character strings is created. This list is sorted alphabetically using the sort() method.

Then, a list of numbers is created. This list is sorted by size using the sort() method. The sort() method is not useful for other types of elements or for mixed lists.

The insert() method is used to insert an element into the list.

The list is reversed internally using the reverse() method.

A specific element is searched for in the list. If the element exists at least once, the first occurrence is deleted via remove(). If the element does not exist, an exception would be triggered. You should therefore check beforehand whether it does exist.

You can append an element to the end of a list by using the append() method.

The number of occurrences of a specific element can be determined using the count() method.

The position of the occurrence of a specific element can be determined using the index() method. You can use a second parameter to specify the position from which the search should start. If the element does not exist in the part of the list currently being searched, an exception will be triggered. You should therefore check beforehand whether it does exist.

With a while loop, you can determine all the positions of a particular element. Your condition is "as long as the element exists in the (current) list." The first time this condition is used, the entire list is checked. Then, only the part of the list after the last position found will be checked.

4.3.5   List Comprehension

The *list comprehension* technique allows you to quickly generate one list from another list. You can filter and change the elements in the first list in the process. Let's look at four examples.

```
x = [3, -6, -8, 9, 15]
print(x)
y = [2, 13, 4, -8, 4]
print(y)
print()

a = [z+1 for z in x]
```

```
print(a)

b = [z+1 for z in x if z>0]
print(b)

c = [x[i]+y[i] for i in range(len(x))]
print(c)

d = [x[i]+y[i] for i in range(len(x)) if x[i]>0 and y[i]>0]
print(d)
```

Listing 4.27 list_comprehension.py File

The output reads:

```
[3, -6, -8, 9, 15]
[2, 13, 4, -8, 4]

[4, -5, -7, 10, 16]
[4, 10, 16]
[5, 7, -4, 1, 19]
[5, 19]
```

First, the two sample lists x and y are created and output.

List a corresponds to list x. However, the value of each element is increased by 1. The expression z+1 for z in x means "Take the elements of the list and add 1."

The situation is similar for list b. However, only the elements with positive values are filtered out. The expression z+1 for z in x if z>0 means "Take the elements of the list and add 1, but only if they are greater than 0."

The index for list c is also included. The elements of the two lists are added in pairs. The expression x[i]+y[i] for i in range(len(x)) means "Add the i-th elements of the two lists over the length of list x." List y could therefore also be larger.

The situation is similar for list d. However, only the pairs with two positive elements are filtered out. The expression x[i]+y[i] for i in range(len(x)) if x [i]>0 and y[i]>0 means "Add the i-th elements of the two lists over the length of list x, but only if both are greater than 0."

4.4 Tuples

A *tuple* corresponds to a list whose elements cannot be changed. Aside from that, the same rules apply, and the same operations and methods can be applied to tuples as to lists as long as they do not cause a change to the tuple.

Let's explore some special features of tuples next.

```
z = (3, 6, -8)
print("Tuple 1 packed:", z)
z = 3, 6, -8
print("Tuple 2 packed:", z)

for i in 3, 6, -8:
    print(i)

a, b, c = z
print("Tuple unpacked:", a, b, c)

a, b, c = 3, 6, -8
print("Multiple assignment:", a, b, c)

a, b, c = c, a, a+b
print("Effect later:", a, b, c)

a, *b, c = 3, 6, 12, -28, -8
print("Rest in list:", a, b, c)
```

Listing 4.28 tuple.py File

Here's the output of the program:

```
Tuple 1 packed: (3, 6, -8)
Tuple 2 packed: (3, 6, -8)
3
6
-8
Tuple unpacked: 3 6 -8
Multiple assignment: 3 6 -8
Effect later: -8 3 9
Rest in list: 3 [6, 12, -28] -8
```

A tuple contains several elements that are separated by commas. They can be written within parentheses. The assignment of a tuple to a single variable is also referred to as *packing* a tuple.

The for loop you're already familiar with uses a tuple.

A tuple can be *unpacked* into multiple variables using an assignment. You must make sure that the number of variables corresponds to the number of elements in the tuple.

You can use a multiple assignment to assign multiple values to multiple variables at the same time. In this case, too, you must make sure you have the right number.

If the assigned values are changed in a multiple assignment, this change will not take effect until the entire assignment is complete. In this example, a is the original value of c, b is the original value of a, and c is the total of the original values of a and b.

A *starred expression* can be used to make a multiple assignment more flexible. Simply place the * operator in front of one of the variables. This operator indicates that the variable refers to a list that contains surplus values after the assignment.

In this example, the first and last values are assigned to variables a and c. The middle values (three in this case) are saved in list b. If only two elements are assigned, list b will remain empty.

4.5   Dictionaries

A *dictionary* or *associative array* can be compared to a glossary. In a dictionary, you'll find the associated information under a key term. For example, in an English-German dictionary, the entry "house" is associated with the German term "Haus."

4.5.1   Properties, Operators, and Methods

In Python, dictionaries represent mutable objects and contain pairs. Each pair consists of a unique key and an assigned value. You can access the value via the key. Strings are usually used as keys, but numbers can also be used. The pairs of a dictionary are unordered.

In the following example, the names and ages of several people are entered and edited in a dictionary. The name serves as a key. This key can be used to access the age information (i.e., the value).

```
dc = {"Peter":31, "Judith":28, "William":35}
print("Dictionary:", dc)
print("Number:", len(dc))
print()

dc_comparison = {"Peter":31, "William":35, "Judith":28}
if dc == dc_comparison:
    print("Equal")

dc["Judith"] = 27
print("Value replaced:", dc)

dc["Noah"] = 22
print("Element added:", dc)
print()
```

```
if "Judith" in dc:
    print(f"Key 'Judith': Value {dc["Judith"]}")
del dc["Judith"]
if "Judith" not in dc:
    print("Key 'Judith' not included")
print("Element removed:", dc)

dc_added = {"Noah": 18, "Kelly": 29}
dc.update(dc_added)
print("Update:", dc)
```

Listing 4.29 dictionary.py File

The following output is generated:

```
Dictionary: {'Peter': 31, 'Judith': 28, 'William': 35}
Number: 3

Equal
Value replaced: {'Peter': 31, 'Judith': 27, 'William': 35}
Element added: {'Peter': 31, 'Judith': 27, 'William': 35, 'Noah': 22}

Key 'Judith': Value 27
Key 'Judith' not included
Element removed: {'Peter': 31, 'William': 35, 'Noah': 22}
Update: {'Peter': 31, 'William': 35, 'Noah': 18, 'Kelly': 29}
```

In this example, a dictionary named dc is generated with three pairs and output. Dictionaries are created using curly brackets. The pairs are separated by commas, and a pair is written in the form *key:value*.

The len() function returns the length (i.e., the number of pairs in the dictionary).

You can only compare dictionaries by using the == and != operators. Two dictionaries match if they contain the same pairs (i.e., the same keys with the same values). Since the pairs of a dictionary are not ordered, the order of the elements is not relevant for the comparison.

You can access an individual value using its key in square brackets. When assigning a pair, the system checks whether the key already exists. If so, only the old value is replaced by the new value. If the key does not already exist, a new pair is added to the dictionary.

You can use the in operator to check whether a specific key exists in the dictionary. This check is often necessary, for example, before outputting a value or before removing a pair using the del operator. An exception occurs when an attempt is made to access a non-existent key.

Since Python 3.12, you can use double quotation marks within an expression in a string literal, as in value {dc["Judith"]}. No conflict exists with the double quotation marks that introduce the string literal.

The update() method is used to update an existing dictionary with the pairs of a new dictionary. When you assign the new pairs, the system checks whether the key already exists. If it exists, the old value is replaced by the new value. If not, a new pair is added to the dictionary.

4.5.2 Dynamic Views

The keys(), items(), and values() methods generate what are called *dynamic views* of a dictionary. If the dictionary changes, these views change as well. These views are used in our next example.

```python
dc = {"Peter":31, "Judith":28, "William":35}
print("Dictionary:", dc)

k = dc.keys()
print("Keys:", k)
for key in k:
    print(key)
if "William" in k:
    print("Key William is included")
print()

v = dc.values()
print("Values:", v)
for value in v:
    print(value)
if 31 in v:
    print("Value 31 is included")
print()

i = dc.items()
print("Items:", i)
for k, v in i:
    print(f"Key {k}, Value {v}")
```

Listing 4.30 dictionary_view.py File

This program generates the following output:

```
Dictionary: {'Peter': 31, 'Judith': 28, 'William': 35}
Keys: dict_keys(['Peter', 'Judith', 'William'])
Peter
```

```
Judith
William
Key William is included

Values: dict_values([31, 28, 35])
31
28
35
Value 31 is included

Items: dict_items([('Peter', 31), ('Judith', 28), ('William', 35)])
Key Peter, Value 31
Key Judith, Value 28
Key William, Value 35
```

The keys() method is used to generate a view of the keys of the dictionary, which is then output. The view has the dict_keys type and is iterable. The individual keys can be output using a for loop. The in operator is used to check whether a specific key exists in the view.

You can use the values() method to generate a view of the values of the dictionary, which is then output. The view has the dict_values type and is iterable. The individual values can be output using a for loop. The in operator is used to check whether a specific value exists in the view.

The items() method is used to generate a view of the key-value pairs of the dictionary, which is then output. The view has the dict_items type and is iterable. The individual pairs can be output using a for loop. In this context, the tuple of key and value is unpacked into two variables.

4.6 Sets

Sets differ from lists and tuples in that each element exists only once.

Sets are unordered, and thus, the order in which a set is output is not fixed. Individual elements cannot be determined on the basis of a slice. However, you can perform some interesting operations with sets that are known from set theory.

4.6.1 Properties, Operators, and Methods

In the next example, a set is created and then edited.

```
st = set([8, 5, 5, 2, 5])
print("Set:", st)
print("Number:", len(st))
```

```
for x in st:
    print(x)
if 5 in st:
    print("Value is included")

su = set([2, 8])
if su < st:
    print("Real subset")
sv = set([2, 8, 5])
if sv <= st:
    print("Subset")

st.add(-12)
st.add(-12)
print("Element added:", st)

st.discard(5)
print("Element removed:", st)

sk = st.copy()
print("Copy:", sk)

sk.clear()
print("Emptied:", sk)

sf = frozenset([8, 5, 5, 2, 5])
print("Frozenset:", sf)
```

Listing 4.31 set.py File

The output reads:

```
Set: {8, 2, 5}
Number: 3
8
2
5
Value is included
Real subset
Subset
Element added: {8, 2, -12, 5}
Element removed: {8, 2, -12}
Copy: {8, 2, -12}
Emptied: set()
Frozenset: frozenset({8, 2, 5})
```

You can create a set using the set() function. The only parameter passed to this function is a list or another iterable. The list transferred in our example contains certain values in multiple places, whereas the set only contains them once.

The len() function returns the length (i.e., the number of elements in a set). You can run through the set using a for loop. By using in, you can check whether a specific element is included in the set.

The four operators, <, <=, >, and >= allow you to determine whether a set is a subset or a proper subset of another set. The order of the elements in the set is irrelevant:

- The su set is a proper subset of the st set: All elements of su are contained in st, and su has fewer elements than st.
- The sv set is a subset of the st set: All elements of sv are contained in st, but sv has at most as many elements as st.

You can use the add() method to add an element if it is not yet included. The discard() method is used to delete a specific element. If you try to delete an element that hasn't been included, no error message arises; the method call is just ignored. The copy() method creates a new set as a copy of an existing set. The clear() method removes all elements from a set.

A *frozenset* is a set that can't be changed. It can be generated using the frozenset() function and identified in the output by the term frozenset. You can run through a frozenset using a for loop and check it using the in operator. Any attempt to change a frozenset raises an exception.

4.6.2 Set Theory

In our next program, we'll use the |, &, -, and ^ operators to perform some operations from mathematical set theory.

```python
st = set([8, 15, "x"])
su = set([4, "x", "abc", 15])
print("st:", st)
print("su:", su)

print("Union:", st | su)
print("Intersection:", st & su)
print("Set difference st - su:", st - su)
print("Set difference su - st:", su - st)
print("Symmetric difference:", st ^ su)
```

Listing 4.32 set_theory.py File

The output reads:

```
st: {8, 'x', 15}
su: {'x', 'abc', 4, 15}
Union: {'x', 4, 8, 'abc', 15}
Intersection: {'x', 15}
Set difference st - su: {8}
Set difference su - st: {'abc', 4}
Symmetric difference: {4, 8, 'abc'}
```

The | operator is used to join two sets. The resulting set contains all the elements that are included in one of the two sets. In the new set, too, each element only appears once.

The elements contained in both sets form the intersection, which can be determined using the & operator.

When dealing with a difference set, you must note which set is subtracted from which other set. The - operator is used to create two different difference sets. The operation st - su subtracts all elements from the st set that are also contained in su. The reverse is true for the su - st operation.

In the symmetric difference set, the ^ operator is used to determine the elements that are contained in only one of the two sets.

4.7   Truth Values and Nothing

Objects and expressions can be true or false, and you also can use the None object. Let's consider a few correlations among these concepts.

4.7.1   Truth Values "True" and "False"

All objects in Python have a truth value: either True or False. In comparisons, such a truth value is determined to control branches and loops (see also Chapter 3, Section 3.3.1). Truth values can be stored in variables of the bool data type. The bool() function returns the truth value of an expression or an object.

The following objects return True:

- A number not equal to 0 (i.e., greater than 0 or less than 0)
- A non-empty sequence (e.g., string, list, or tuple)
- A non-empty dictionary
- A non-empty set

The following objects return False:

- A number that has the value 0
- An empty sequence (i.e., the string "", the list [], or the tuple ())

- An empty dictionary: {}
- An empty set, either set() or frozenset()
- A collection of objects x for which the following holds true: len(x) == 0
- The None constant (Section 4.7.2)

For a list, tuple, string, dictionary, or set, as long as the object collection contains elements, the truth value is True. If the elements are removed, the truth value is False. In the next program, the truth values of various objects are checked and output.

```
print("5>3:", 5>3)
print("5<3:", 5<3)
print("Type of 5>3:", type(5>3))

print("Number -0.1:", bool(-0.1))
print("Number 0:", bool(0))

print("String 'Hello':", bool("Hello"))
print("String '':", bool(""))

print("List [2,8]:", bool([2,8]))
print("List []:", bool([]))

print("Tuple (2,8):", bool((2,8)))
print("Tuple ():", bool(()))

print("Dictionary {'Judith':28, 'William':32}:",
      bool({"Judith":28, "William":32}))
print("Dictionary {}:", bool({}))

print("Set (2,8,2):", bool(set([2,8,2])))
print("Set ():", bool(set([])))
```

Listing 4.33 truth_value.py File

The program generates the following output:

```
5>3: True
5<3: False
Type of 5>3: <class 'bool'>
Number -0.1: True
Number 0: False
String 'Hello': True
String '': False
List [2,8]: True
List []: False
```

```
Tuple (2,8): True
Tuple (): False
Dictionary {'Judith':28, 'William':32}: True
Dictionary {}: False
Set (2,8,2): True
Set (): False
```

The result for string, list, tuple, dictionary, and set is `False` if empty and `True` if not empty. You can use this function to check the objects in question.

4.7.2   Nothing and the None Keyword

The `None` keyword denotes an object without content. `None` is the only object of the `NoneType` data type. Functions without a return value return `None`. This value may be an indication of the following:

- That you've used a function incorrectly where you expect a return value
- That a function does not return a result although you expect one

Let's consider an example.

```
def total(a, b):
    c = a + b

# Program
res = total(7,12)
if not res:
    print("Error")
if res is None:
    print("Error")
print("Result:", res)
print("Truth value:", bool(res))
print("Type:", type(res))
```

Listing 4.34 nothing.py File

The output reads:

```
Error
Error
Result: None
Truth value: False
Type: <class 'NoneType'>
```

When defining the total() function, the return of the result is "forgotten." For this reason, res receives the value None. The truth value of the nothing object is False. In this case, you can use this function to determine if an error exists.

4.8   Reference, Identity, and Copy

In this section, I want to explain the relationship between objects and references. We'll also examine the identity of objects and create copies of objects.

4.8.1   Reference and Identity

The name of an object is basically a reference to the object.

If you assign this reference to a different name, you create a second reference to the same object. You can use the is identity operator to determine that both references refer to the same object.

If the content of the object is changed via the second reference, one of the following two behaviors occurs:

- In the case of a simple object, such as a number or string, a second object is created in which the new value is saved. The two references then refer to two different objects.

- In the case of a non-simple object such as a list, tuple, dictionary, set, etc., the original object is changed. There is still one object with two different references.

You can use the == operator to determine whether two objects have the same content, for example, whether two lists contain the same elements.

In the following example, a number, a string, and a list are created one after the other and referenced twice. New content is then assigned to the second reference. Identity and content are determined using the is and == operators.

```
print("Number:")
x = 12.5
y = x
print("same object:", x is y)
y = 15.8
print("same object:", x is y)
print("same content:", x == y)
print()

print("String:")
x = "Robinson"
y = x
print("same object:", x is y)
```

```
y = "Friday"
print("same object:", x is y)
print("same content:", x == y)
print()

print("List:")
x = [23,"hello",-7.5]
y = x
print("same object:", x is y)
y[1] = "world"
print("same object:", x is y)
print("same content:", x == y)

y = ["abc",12.8,"xyz"]
print("same object:", x is y)
```

Listing 4.35 reference.py File

The following output is generated:

```
Number:
same object: True
same object: False
same content: False

String:
same object: True
same object: False
same content: False

List:
same object: True
same object: True
same content: True
same object: False
```

The output shows that the objects are initially identical.

Assigning a new value to a number or string creates a new object. The contents are then different.

The list (in this case, also representing other objects) only exists once in total, even if individual elements of the list are changed. You can access the same list via both references.

A second object is only created after a new list has been assigned.

4.8.2   Saving Resources

Python likes to save resources, which can lead to unusual behaviors: If a value is assigned to an object via a reference *and* the same value is already referenced by another reference, it can happen that the two references subsequently refer to the same object. In this way, Python saves memory space.

The del keyword can be used to delete references that are no longer required. An object pointed to by two references does not get deleted by deleting the first reference.

Let's consider an example.

```
x = 42
y = 56
print(f"x:{x}, y:{y}, same object: {x is y}")

y = 42
print(f"x:{x}, y:{y}, same object: {x is y}")

del y
print("x:", x)

del x
try:
    print("x:", x)
except:
    print("Error")
```

Listing 4.36 resources.py File

The following output is generated:

```
x:42, y:56, same object: False
x:42, y:42, same object: True
x: 42
Error
```

First, the x and y variables are given different values. They are therefore two references to two different objects. The y variable is then given the same value as the x variable. Now, there is only one object that is pointed to by two references.

The y reference is deleted, while the object still exists and can be accessed via the x reference. Then, the x reference is deleted. The output via this reference leads to an error because the reference no longer exists.

4.8.3 Copying Objects

You can create real copies of non-simple objects such as lists, tuples, dictionaries, sets, and so on by creating an empty object and adding the individual elements.

This process can be rather time consuming for large objects that in turn contain other objects. For this reason, we recommend using the deepcopy() function from the copy module.

Both procedures are shown in our next program.

```
import copy

x = [23,"hello",-7.5]
y = []
for i in x:
    y.append(i)
print("same object:", x is y)
print("same content:", x == y)
print()

x = [23,["Boston","Houston"],-7.5,12,67]
y = copy.deepcopy(x)
print("same object:", x is y)
print("same content:", x == y)
```

Listing 4.37 copy_object.py File

The program generates the following output:

```
same object: False
same content: True

same object: False
same content: True
```

The output shows that a new object is created in both cases. However, the contents of the two objects are the same.

4.9 Type Hints

In Chapter 3, Section 3.8, I introduced the concept of type hints and type checks. Type hints can improve the readability of the code, and type checks can make it easier to search for errors. This section deals with the connection between type hints and sequences.

4.9.1   A Sequence as a Parameter

The generic last() function from our next program returns the last element of a sequence.

```
def last(li):
    return li[-1]

print("String:", last("Good morning"))
print("Tuple:", last((3, -14, 12)))
print("List:", last([7, 2, 16, -5, 52, 2]))
```

Listing 4.38 type_hints_sequence.py File, Version 1

This program is called using three different sequences that can be indexed: a string, a tuple, and a list, with the following result:

```
String: g
Tuple: 12
List: 2
```

4.9.2   Restriction to a List

Type hints can be used to specify that a function only allows a list as a parameter and no other sequence. For this purpose, let's change our program.

```
def last[T](li:list[T]) -> T:
    return li[-1]

print("String:", last("Good morning"))
print("Tuple:", last((3, -14, 12)))
print("List:", last([7, 2, 16, -5, 52, 2]))
```

Listing 4.39 type_hints_sequence.py File, Version 2

The name of the function is followed by an identifier in square brackets, which acts as a placeholder for a type. By convention, the name T is used. The parameter is a list of elements of type T. The return value is a single value of type T. The result of the program remains the same. A type check using one of the *pyright* or *mypy* programs returns the two error messages shown in Figure 4.2.

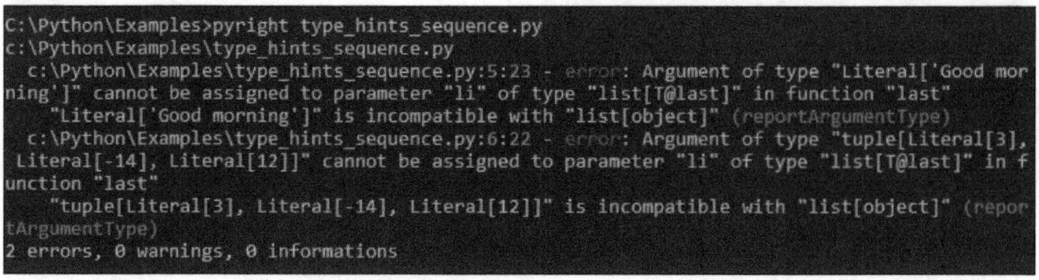

```
C:\Python\Examples>pyright type_hints_sequence.py
c:\Python\Examples\type_hints_sequence.py
   c:\Python\Examples\type_hints_sequence.py:5:23 - error: Argument of type "Literal['Good mor
ning']" cannot be assigned to parameter "li" of type "list[T@last]" in function "last"
      "Literal['Good morning']" is incompatible with "list[object]" (reportArgumentType)
   c:\Python\Examples\type_hints_sequence.py:6:22 - error: Argument of type "tuple[Literal[3],
 Literal[-14], Literal[12]]" cannot be assigned to parameter "li" of type "list[T@last]" in f
unction "last"
      "tuple[Literal[3], Literal[-14], Literal[12]]" is incompatible with "list[object]" (repor
tArgumentType)
2 errors, 0 warnings, 0 informations
```

Figure 4.2 Incorrect Sequence Type

Chapter 5
Advanced Programming

In this chapter, the knowledge you've gained from our programming course is expanded with useful practical tips, especially in connection with the various data types.

5.1   General Remarks

In the first part of this chapter, I explain some generally useful techniques that cannot be easily categorized.

5.1.1   Combined Assignment Operators

In addition to the simple assignment of a value to a variable, you can also use *combined assignment operators*. These operators combine the normal operators for numbers or strings with the assignment of a value. In this way, the respective variable is immediately changed by the specified value, which is particularly useful for extensive expressions or longer variable names.

Let's consider some examples: The TemperatureInCelsius += 5 expression is more straightforward than the TemperatureInCelsius = TemperatureInCelsius + 5 expression. Both expressions increase the value of the TemperatureInCelsius variable by 5.

Combined assignment operators can be used for numbers and character strings, as illustrated in our next example.

```
x = 12
print(x)
x += 3      # Increasing x
print(x)
x -= 9      # Decreasing x
print(x)
x **= 2     # Squaring x
print(x)
x *= 3      # Tripling x
print(x)
x //= 7     # Integer division of x
```

```
print(x)
x /= 4        # Dividing x
print(x)
x %= 2        # Dividing, calculating the remainder
print(x)

t = "Hello "
print(t)
t += "Python "    # Append to t
print(t)
t *= 3            # Tripling t
print(t)
```

Listing 5.1 assignment_combined.py File

The following output is generated:

```
12
15
6
36
108
15
3.75
1.75
Hello
Hello Python
Hello Python Hello Python Hello Python
```

The x variable is initially assigned the numerical value 12. The subsequent statements change its value each time. The value is first increased by 3 (= 15), then decreased by 9 (= 6), then raised to the power of 2 (= 36), and multiplied by 3 (= 108).

These steps are followed by an integer division in which the decimal places are removed. The remaining number (15) is divided by 4 (= 3.75). Finally, the remainder of the division by 2 is calculated (= 1.75).

The t variable receives the string value Hello. The variable is then extended and multiplied.

When using combined assignments, you must ensure that the relevant variable already has a value beforehand; otherwise, an error will occur.

5.1.2 Statement in Multiple Lines

In Chapter 2, Section 2.3.6, I described how a long character string might run across multiple lines when using the print() function. In this section, I will show you how you can split up long statements in general, using the \ character, among other things. Let's look at an example:

```
print("Please enter a"
      " temperature in Celsius:")
TemperatureInCelsius = float(input())

TemperatureInFahrenheit = TemperatureInCelsius \
                    * 9 / 5 + 32

print(TemperatureInCelsius, "degrees Celsius correspond to",
      TemperatureInFahrenheit, "degrees Fahrenheit")
```

Listing 5.2 line_long.py File

The following output is generated:

```
Please enter a temperature in Celsius:
5.2
5.2 degrees Celsius correspond to 41.36 degrees Fahrenheit
```

First, a longer string is split across two lines. Each part of the string is placed in double quotation marks. A separating space between individual parts must be entered manually.

The \ character is used for a longer statement that must be separated within a calculation. It indicates that the statement will continue in the subsequent line.

You can easily separate a statement containing commas after one of the commas.

5.1.3 Input with Assistance

The built-in input() function for entering strings has an optional parameter that allows you to transfer useful information for input. In our next example, the names of three cities should be input and saved in a list.

```
li = []
for i in range(3):
    tx = input(f"No {i+1}: Please enter a city: ")
    li.append(tx)
print(li)
```

Listing 5.3 input_help.py File

The following output is generated:

```
No 1: Please enter a city: Paris
No 2: Please enter a city: Rome
```

```
No 3: Please enter a city: Madrid
['Paris', 'Rome', 'Madrid']
```

The consecutive number of the city to be entered is displayed as an aid during input.

5.1.4 The pass Statement

The pass statement indicates that nothing will be executed. Some possible areas of use for this statement include the following scenarios:

- You develop a program in which a function is called that will be developed at a later date. First, the main program needs to be written, which also contains a call of the function. The definition of the function initially contains only the pass statement.
- The program contains a branch in which nothing is to be executed in a specific branch. However, this branch should still appear in order to make the program flow clearer.

Let's look at a program with examples of these cases next.

```
def SquaringTheCircle():
    pass
print("The next step is to square a circle")
SquaringTheCircle()
print("The circle has been squared")

a = -12
b = 6
c = 6.2
if a >= 0 and b >= 0 and c >= 0:
    pass
else:
    print("One of the numbers is negative")

if a == 1:
    print("Case 1")
elif a == 2:
    print("Case 2")
elif a < 0:
    pass
else:
    print("Otherwise")
```

Listing 5.4 pass.py File

The following output is generated:

```
The next step is to square a circle
The circle has been squared
One of the numbers is negative
```

The `SquaringTheCircle()` function initially serves only as a dummy and will be filled with content at a later stage. However, we've already built it into the program, and it can be called.

If you use the simple branch, an output will only occur in the `else` branch. If you use the multiple branch, an output will only occur if the examined value is greater than or equal to 0.

5.2   Output and Formatting

In this section, I will explain some options for output using the `print()` function and for formatting outputs.

5.2.1   The print() Function

We've already used the `print()` function frequently, but it offers even more possibilities:

- The *separator* that separates the output objects from each other can be changed using the named sep parameter. This parameter contains a space by default.
- The end-of-line character, which follows an output by default, can be changed using the named end parameter.

Let's consider some examples.

```
a = 23
b = 7.5
c = a + b
print("Result: ", end="")
print(a, "+", b, "=", c, sep="")

li = ["Houston", "Boston", "Austin"]
for city in li:
    print("City", city, sep="=>", end=" # ")
```

Listing 5.5 print.py File

The following output is generated:

```
Result: 23+7.5=30.5
City=>Houston # City=>Boston # City=>Austin #
```

Let's first take a look at the output of the calculation. Two calls of the `print()` function normally generate two output lines, but a difference exists:

- The end parameter is set to an empty character string for the first call. As a result, the next output will be in the same line.

- In the second call, the sep parameter is set to an empty character string. As a result, the individual parts of the output are joined together directly, without spaces.

Both the separator and the end-of-line character have been changed for the output of the list.

5.2.2   Formatting Numbers with Decimal Places

String literals have been available since Python 3.6. We've already used them frequently for the convenient embedding of expressions in character strings.

In this section and in the next two sections, Section 5.2.3 and Section 5.2.4, I describe additional possibilities. For example, you can clearly display the columns of tables in a standardized formatting. For this task, you can specify, among other things, the minimum output width of the numbers and the number of decimal places to which they are rounded.

These formatting options explain the full name of the string literals: They are technically *formatted string literals*, or *f-strings* for short. Since Python 3.12, string literals are processed internally for the most part like real components of Python statements. For this reason, string literals are more flexible, faster, and more suitable for a wide range of development environments. In addition, error messages provide more information.

In our next program, some numbers with decimal places are output using formatted string literals.

```
x = 100/7
y = 2/7
print("Numbers:", x, y)
print()

print(f"Format f, Standard:           {x:f}")
print(f"Format f, after decimal point: {x:.25f}")
print(f"Format f, total:              {x:15.10f}")
print()

print(f"Format e, Standard:           {x:e}")
```

```
print(f"Format e, after decimal point: {x:.3e}")
print(f"Format e, total:               {x:12.3e}")
print()

print(f"Format %, Standard:            {y:%}")
print(f"Format %, after decimal point: {y:.3%}")
print(f"Format %, total:               {y:12.3%}")
```

Listing 5.6 literal_decimal_place.py File

The expected output is shown in Figure 5.1.

```
Numbers: 14.285714285714286 0.2857142857142857

Format f, Standard:            14.285714
Format f, after decimal point: 14.285714285714286475581502
Format f, total:                 14.2857142857

Format e, Standard:            1.428571e+01
Format e, after decimal point: 1.429e+01
Format e, total:                  1.429e+01

Format %, Standard:            28.571429%
Format %, after decimal point: 28.571%
Format %, total:                  28.571%
```

Figure 5.1 Formatting Numbers with Decimal Places

The x and y variables are given the values 100/7 and 2/7, respectively. They are output unformatted for comparison reasons.

As usual, a string literal is created using the f character and the curly brackets. However, the element to be embedded is now followed by a colon and the formatting characters.

The formatting character f represents the output of a number with a certain number of decimal places, to which it is rounded (by default, 6 decimal places). The .25f specification generates 25 decimal places after the decimal point. The 15.10f specification represents the right-aligned output of a number with a minimum total width of 15 places, 10 of which are after the decimal point. Such a format is particularly suitable for tables.

The formatting character e represents the output of a number in exponential format (by default, 6 decimal places and an exponent). The .3e specification generates 3 decimal places. The 12.3e specification stands for a right-aligned output with a minimum total width of 12 characters, 3 of which are after the decimal point.

The formatting character % represents the output of a number in percentage format (by default, with 6 decimal places and a percent sign). The number is multiplied by 100 (only for the output), whereas internally it remains unchanged. The .3% specification generates 3 places after the decimal point. The 12.3% specification stands for a right-aligned output with a minimum total width of 12 places, 3 of which are after the decimal point.

Note

Pay attention to spelling when formatting. For example, you cannot have a space before the closing curly bracket.

5.2.3 Formatting Integers

The following is an output of integers using formatted string literals:

```
for z in range(61,67):
    print(f"{z:4d}{z:9b}{z:4o}{z:4x}")
```

Listing 5.7 literal_integer.py File

The expected output is shown in Figure 5.2.

```
61   111101   75  3d
62   111110   76  3e
63   111111   77  3f
64  1000000  100  40
65  1000001  101  41
66  1000010  102  42
```

Figure 5.2 Formatting of Integers

A uniformly formatted table of numbers is generated. The numbers from 61 to 66 are output one after the other in several formats:

- Decimal number (character d, with the minimum total width of 4)
- Dual number (character b, with the minimum total width of 9)
- Octal number (character o, with the minimum total width of 4)
- Hexadecimal number (character x, with the minimum total width of 4)

5.2.4 Formatting Strings

In the following table, the output of strings is added using formatted string literals.

```
itemname = {23: "Apple", 8: "Banana", 42: "Peach"}
quantity = {23:1, 8:3, 42:5}
uprice = {23:2.95, 8:1.45, 42:3.05}

print(f"{'No':>4}{'Name':>12}{'Quantity':>9}{'UP':>10}{'TP':>10}")
for x in 23, 8, 42:
    print(f"{x:04d}{itemname[x]:>12}{quantity[x]:9d}"
          f"{uprice[x]:8.2f} ${quantity[x] * uprice[x]:8.2f} $")
```

Listing 5.8 literal_string.py File

The expected output is shown in Figure 5.3.

```
 No          Name Quantity         UP         TP
0023        Apple        1     2.95 $     2.95 $
0008       Banana        3     1.45 $     4.35 $
0042        Peach        5     3.05 $    15.25 $
```

Figure 5.3 Formatting of Strings

You can use formatting options not only for variables or calculation expressions, but also for values, as we did here for the character strings in the heading of the items table. A string within a string is written in single quotation marks.

By default, numbers are right aligned, and strings are left aligned. The formatting character > represents a right-aligned output, while the number that follows it stands for the minimum total width of the output. The < character stands for left aligned.

The items table is based on the three dictionaries for the item name, the quantity, and the unit price. The elements of the dictionaries are output uniformly along with the calculated total price.

If you have a zero before the total width, the number is filled with leading zeros, as is the case for the first column. When splitting a single string literal into multiple string literals, you must note the leading f before each string.

5.2.5   Exercise

u_literal

Rewrite the program from the u_range_inch exercise (from Chapter 3, Section 3.4.7) in such a way that the output shown in Figure 5.4 is generated. The values for the columns "inch" and "cm" should each be displayed with one decimal place and right aligned (*u_literal.py* file).

```
inch      cm
15.0    38.1
20.0    50.8
25.0    63.5
30.0    76.2
35.0    88.9
40.0   101.6
```

Figure 5.4 u_literal

5.3   Functions for Iterables

A number of functions work with iterables. These functions can take a lot of work off your hands at development time. As examples, I will focus on the zip(), map(), and filter() functions in this section.

Python is a rather versatile programming language. It follows the principles of functional programming when using the following functions and technology:

- The map() function (Section 5.3.2)
- The filter() function (Section 5.3.3)
- The *list comprehension* (see Chapter 4, Section 4.3.5)
- The *lambda function* (Section 5.6.8)

5.3.1 The zip() Function

The zip() function returns an object of the zip type. Similar to a zipper, the contents of iterables are linked together in such an object. An object of the zip type is itself iterable. Let's consider an example.

```
name = ["Mayer", "Smith", "Mertins"]
firstname = ["John", "Peter", "Judith"]
personnelnumber = [6714, 81343, 2297]
salary = [3500.0, 3750.0, 3621.5]
birthday = ["03/15/1962", "04/12/1958", "12/30/1959"]

combi = zip(name, firstname, personnelnumber, salary, birthday)
for element in combi:
    print(element)
print(type(combi))
```

Listing 5.9 iterable_zip.py File

The following output is generated:

```
('Mayer', 'John', 6714, 3500.0, '03/15/1962')
('Smith', 'Peter', 81343, 3750.0, '04/12/1958')
('Mertins', 'Judith', 2297, 3621.5, '12/30/1959')
<class 'zip'>
```

First, various iterables are created—in this case, five lists containing data of different people: family name, first name, personnel number, salary, and date of birth. The zip() function receives the five iterables as parameters. The function returns an object of type zip. The data of the five iterables are linked together in this object. The object consists of individual tuples and is output using a for loop.

5.3.2 The map() Function

The map() function allows you to call your own function multiple times in one statement, each time with different parameters. This function returns an object of the map type and consists of the results of the calls of its own function and can be iterated.

Our next program shows two examples of using the map() function.

```
def quad(x):
    return x * x

def add(a,b,c):
    return a + b + c

z = map(quad, [4, 2.5, -1.5])
print("Square:")
for element in z:
    print(element)
print()

z = map(add, [3, 1.2, 2], [4.8, 2], [5, 0.1, 9])
print("Total:")
for element in z:
    print(f"{element:.1f}")
print()

print(type(z))
```

Listing 5.10 iterable_map.py File

The following output is generated:

```
Square:
16
6.25
2.25

Total:
12.8
3.3

<class 'map'>
```

First, two separate functions are defined:

- The quad() function has one parameter and returns its square.
- The add() function has three parameters and returns their sum total.

The following information is transferred when the map() function is first called:

- The first parameter is the name of the quad() function.
- The second parameter is an iterable containing the values for the various calls of the quad() function.

- The z object of type map is returned. This object contains the return values of the quad() function for the various calls.
- These values are output using a for loop.

The second call of the map() function passes the following information:

- The first parameter is the name of the add() function.
- The second and all subsequent parameters are iterables containing the values for the various calls of the function.
- The z object of type map is returned. This object contains the return values of the add() function for the various calls.
- These values are output using a for loop.
- To form the first total, the first element of each iterable (in our case, 3, 4.8, and 5) is used. To form the second total, the second element of each iterable (in our case, 1.2, 2, and 0.1) is used, and so on.
- The shortest iterable determines the number of calls.

5.3.3 The filter() Function

The filter() function examines elements of an iterable using another function. It returns an object of the filter type. This object can be iterated and contains the elements for which the other function returns the truth value True.

Let's look at an example:

```
def positive(a):
    if a>0:
        return True
    else:
        return False

z = filter(positive, [5, -6, -2, 0, 12, 3, -5])
for element in z:
    print(element)
print(type(z))
```

Listing 5.11 iterable_filter.py File

The following output is generated:

```
5
12
3
<class 'filter'>
```

The first parameter of the filter() function is the name of the other function, which returns True or False for an examined value. The second parameter is the iterable. An object of the filter type is returned, which contains the elements for which the other function returns True.

5.4 Encryption

The functions of the random module are sufficient for the random values of a game. However, if you need random values for the purposes of encryption or for security reasons, you should use the functions of the secrets module, which provides secure random numbers and has been available since Python 3.6.

Let's consider a simple example.

```python
import string, secrets

li = []
for i in range(10):
    li.append(secrets.randbelow(6) + 1)
print("Roll the dice ten times:", li)

tx = ""
for i in range(10):
    tx += secrets.choice(string.ascii_lowercase)
print("Text with ten random lowercase letters:", tx)
```

Listing 5.12 secrets_functions.py File

The randbelow() function returns a random number between 0 and the value of the transferred parameter. The die is rolled ten times.

The choice() function returns a random value from a sequence. The sequence in this example is the ascii_lowercase string constant from the string module, which contains the lowercase letters from the American Standard Code for Information Interchange (ASCII) code (see also Chapter 4, Section 4.2.8). A sequence of ten random lowercase letters is determined and output.

A possible output of the program could read as follows:

```
Roll the dice ten times: [2, 4, 3, 4, 3, 2, 1, 1, 5, 4]
Text with ten random lowercase letters: dxhrankfme
```

In the following program, the choice() function is used to create a random password consisting of six characters. The password must contain at least one character from each of the following four groups: lowercase letters, uppercase letters, numbers, and special characters.

```python
import string, secrets

special = "!#$%&()*+-/:;<=>?@[]_{|}"
print("Special characters:", special)
ch_all = string.ascii_letters + string.digits + special

while True:
    tx = ""
    num = [0,0,0,0]
    for i in range(6):
        character = secrets.choice(ch_all)
        tx += character
        if character in string.ascii_lowercase:
            num[0] += 1
        elif character in string.ascii_uppercase:
            num[1] += 1
        elif character in string.digits:
            num[2] += 1
        else:
            num[3] += 1
    print("Number:", num)
    if 0 not in num:
        break

print("Password: ", tx)
```

Listing 5.13 secrets_password.py File

First, a character string is created that consists of all special characters that are permitted in this case. Next, a character string is put together that consists of all the characters permitted for the password (i.e., all uppercase and lowercase letters, all numbers, and all possible special characters).

The password is created in an endless loop that is exited once the password has fulfilled the conditions mentioned previously. For the password, the choice() function is used to take six characters in succession from the set of permitted characters and put them together.

We also have a list of counters, one counter for each group. The group membership is determined for each character. The corresponding counter is incremented. When no more counter shows 0, the infinite loop will be exited.

In the following possible outputs of the program, you can see how several runs of the loop were necessary:

```
Special characters: !#$%&()*+-/:;<=>?@[]_{|}
Number: [1, 3, 0, 2]
Number: [3, 2, 0, 1]
Number: [2, 2, 1, 1]
Password:  B(kO7w
```

5.5   Errors and Exceptions

This section provides further explanations and techniques related to errors and exceptions.

5.5.1   General Remarks

While you're developing and testing a program, errors will occur. This is normal and also important for the learning process. There are three types of errors:

- You'll notice *syntax errors*, at the latest, when you start a program.

- *Runtime errors*, that is, errors at program runtime that cause the program to crash, can be handled using a try-except block.

- *Logical errors* occur if the program runs completely but does not deliver the expected results. The process was not properly thought through in this case. Experience has shown that these errors are the most difficult to find. Debugging will be useful in this context.

5.5.2   Syntax Errors

Syntax errors occur at program development time and are caused by incorrectly or incompletely written program code. Python draws attention to syntax errors, at the latest, when a program is started by displaying a message and a reference to the error location. Then, the program will no longer be executed.

Let's look at an example of some incorrect code.

```
x = 12
if x > 10
    print(x
```

Listing 5.14 error_code.py File

After starting the program, an error message appears because a colon is expected. In Python 3.11 and Python 3.12, the analysis of errors has been greatly improved. The resulting messages are more accurate, as shown in Figure 5.5.

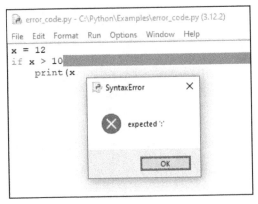

Figure 5.5 Display of the First Error

The program is no longer running. It cannot be restarted until the program has been improved (i.e., after the colon has been inserted). Another error message appears because the parentheses of the print() function were not closed, as shown in Figure 5.6. The program only runs to the end after the last error has been eliminated.

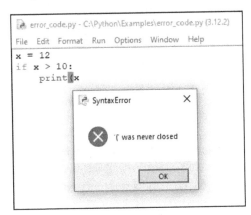

Figure 5.6 Display of the Second Error

5.5.3   Runtime Errors

You can use a try-except block to catch runtime errors that lead to exceptions, as described in Chapter 3, Section 3.6. Runtime errors occur if the program attempts to perform an invalid operation, for example, a division by 0 or the opening of a non-existent file.

You cannot completely avoid runtime errors, as there are processes over which you have no influence during development. Errors may arise from incorrect inputs during use or the absence of a necessary file with input data. I will explain further options for intercepting runtime errors in Section 5.5.6.

5.5.4 Logical Errors and Debugging

Logical errors occur if an application is translated without syntax errors and executed without runtime errors but does not deliver the intended result. This error arises due to an error in the program logic.

Finding the cause of logical errors is often difficult and requires intensive testing and analysis of the processes and their results. The Integrated Development and Learning Environment (IDLE) provides a simple debugger for this purpose.

Single-Step Mode

In the debugger, you can run a program in single-step mode and view the current contents of variables for each of these individual steps. Let's use a simple program with a loop and a function to serve as an example.

```
def total(a,b):
    c = a + b
    return c

for i in range(5):
    res = total(10,i)
    print(res)
```

Listing 5.15 error_debugging.py File

This program writes the numbers from 10 to 14 on the screen one after the other. First, open the **IDLE Shell** and add this program into the window.

You can start the debugger by calling the **Debugger** menu item in the **IDLE Shell** within the **Debug** menu. The **Debug Control** dialog box appears, and the [DEBUG ON] message is displayed in the **IDLE Shell**, as shown in Figure 5.7.

Now, start the program as usual in the program window via the **Run • Run Module** menu path or press F5. The **Debug Control** dialog box refers to the first line of the program, as shown in Figure 5.8.

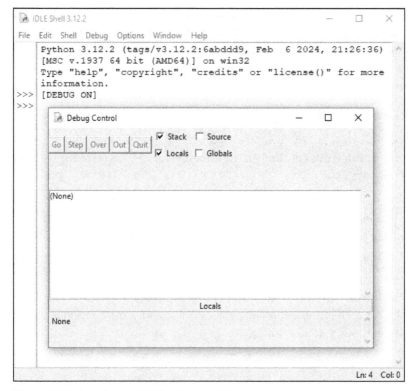

Figure 5.7 Debug Control Dialog Box and Message

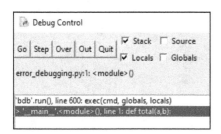

Figure 5.8 First Executed Program Line

You can now use the buttons in the **Debug Control** dialog box. The **Step** button allows you to go through the program step by step. The next step takes you directly to the first executed program line, behind the function definition.

You can now run through the loop several times by repeatedly clicking on the **Step** button. The total() function is run each time. In the lower area of the **Debug Control** dialog box, you can see the variables and their constantly changing values. If you're in the loop during one of the steps in the main program, you'll see the values of i and res, as shown in Figure 5.9.

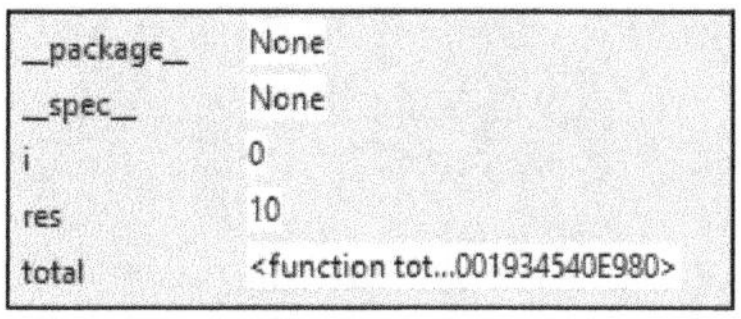

package	None
spec	None
i	0
res	10
total	<function tot...001934540E980>

Figure 5.9 Main Program: Current Values of i and res

If you're in one of the steps within the total() function, you'll see the values of a, b, and c, as shown in Figure 5.10.

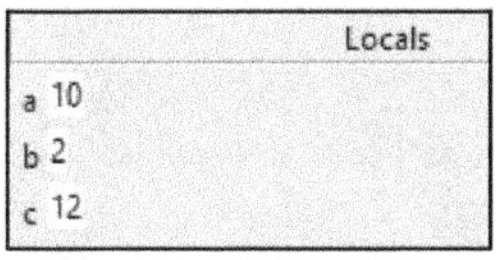

Locals	
a	10
b	2
c	12

Figure 5.10 Function: Current Values of a, b, and c

The first results are displayed in parallel in the **IDLE Shell**, as shown in Figure 5.11.

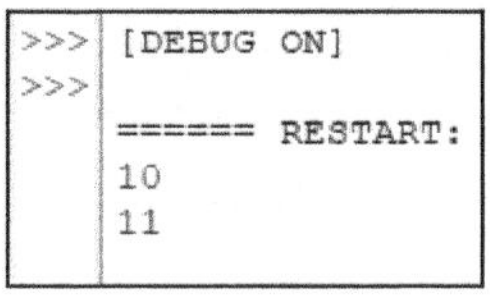

```
>>> [DEBUG ON]
>>>
       ====== RESTART:
    10
    11
```

Figure 5.11 Output of the First Results in the IDLE Shell

After running through all program lines step by step, the [DEBUG ON] message is displayed in the **IDLE Shell** once more. IDLE is still in debug mode, but you cannot use the buttons again until you have restarted the program.

Additional Options

To run through the program in somewhat larger steps, click the **Over** button. In this case, the functions are not run through in individual steps, but in their entirety, which means that the debugger jumps over the functions.

You can also switch back and forth between the two options (**Step** and **Over**), depending on which part of the program you want to view in detail.

If you're currently moving through a function in individual steps, clicking the **Out** button will skip the rest of the function and continue with the first step after the function call.

The **Go** button runs the program you're currently debugging to the end in one step. The **Quit** button terminates the program immediately without allowing it to run to the end. In both cases, the debug mode is still switched on.

Debug mode can be switched off in the same way as you switched it on: In the **IDLE Shell** after calling up the menu item **Debug • Debugger**. The [DEBUG OFF] message then appears in the **IDLE Shell**.

A less elegant way to exit the debugger is to simply close the **Debug Control** dialog box.

Other development environments for Python provide additional options. Setting *breakpoints* is especially useful. These breakpoints are set to specific program lines. The program then runs up to such a program line in one go, and you can check the current values. You then either run through a program area in which you suspect errors using the single-step procedure, or you go directly to the next breakpoint that has already been set, and so on.

5.5.5   Generating Errors

"Why would I want to generate errors?" you may ask yourself after reading this heading. Quite sensible reasons for creating errors exist, especially in connection with entering data while using the program.

In our next example, a number is to be entered whose reciprocal value is then calculated by the program. However, the number you enter should be positive. This restriction can be edited using the raise statement, which is used to generate an error.

```
while True:
    try:
        number = float(input("A positive number: "))
        if number < 0:
            raise
        rp = 1.0 / number
        break
    except:
        print("Error")

print(f"The reciprocal of {number} is {rp}")
```

Listing 5.16 error_generate.py File

If the number you enter is less than 0, the raise statement will be executed, which creates an exception. The program then goes directly to the except branch and executes the statements specified there.

In this case, we only have a logical input error, but this error is treated in the same way as any error in the program. This error ensures that only the reciprocal of a positive number will be calculated. Consider the following possible inputs:

```
A positive number: 0
Errors
A positive number: abc
Error
A positive number: -6
Error
A positive number: 6
The reciprocal of 6.0 is 0.16666666666666666
```

Various mistakes are made in these examples, such as the following:

- The number 0 is entered. This input would lead to an exception of the ZeroDivisionError type during the calculation of the reciprocal value.

- A text is entered. This input would lead to an exception of the ValueError type when the float() function is called.

- A negative number is entered. This input leads to a self-generated exception due to the restriction made.

The try, raise, and except statements can therefore be used to intercept and handle inputs that are permitted in Python but are not useful for the course of the program. However, the technique demonstrated in this section still has a disadvantage in that all errors are treated equally, and therefore, the resulting error messages are not yet perfectly accurate. This problem will be improved in the following section.

5.5.6   Distinguishing Exceptions

In the next program, different types of errors that cause exceptions are specifically intercepted. Thus, more precise error messages will appear, and the program is easier to use. Again, the reciprocal value of an entered number is to be determined.

```python
while True:
    try:
        number = float(input("A positive number: "))
        if number == 0:
            raise RuntimeError("Number equals 0")
        if number < 0:
            raise RuntimeError("Number too small")
        rp = 1.0 / number
        break
    except ValueError:
        print("Error: no number")
    except ZeroDivisionError:
        print("Error: number 0 entered")
```

```
    except RuntimeError as e:
        print("Error:, e)

print(f"The reciprocal of {number} is {rp}")
```

Listing 5.17 error_distinguish.py File

The program contains several specific interception options for a test. Consider the following possible inputs:

```
A positive number: 0
Error: Number equals 0
A positive number: abc
Error: no number
A positive number: -6
Error: Number too small
A positive number: 6
The reciprocal of 6.0 is 0.16666666666666666
```

Again, various mistakes are made in these examples, such as the following:

- The number 0 is entered. This input leads to a runtime error due to the restriction made. This error is intercepted as a general RuntimeError with the Error message Number equals 0.

- If the input of 0 were not intercepted in this way, a runtime error would occur later when calculating the reciprocal value, a ZeroDivisionError. This error would be intercepted with the message Error: Number 0 entered. The error is intercepted twice for demonstration purposes.

- A text is entered. This input leads to a runtime error when the float() function is called, namely, a ValueError, which in turn is intercepted with the message Error: no number.

- A negative number is entered. This input causes a runtime error due to the second restriction. The error is also intercepted as a general RuntimeError with the message Error: Number too small.

When the error is generated using the raise statement, an error and a message are transferred. When intercepting an error using the except statement, the message is transferred to the e variable using the as keyword.

5.6   Functions

Python provides some rather useful extensions on the subject of functions, which I will explain in this section.

5.6.1 A Variable Number of Parameters

So far, we have made sure that the number of parameters of a function is the same when defining and calling it. By using a *starred expression* (see Chapter 4, Section 4.4), you can also define functions with a variable number of parameters.

Place the * operator in front of a parameter of the function. This parameter therefore refers to a tuple that contains the surplus values of the parameter chain after the function is called.

In the next example, a function is defined and called twice to calculate and return the total of all parameters:

```python
def total(a, *summands):
    res = a
    for s in summands:
        res += s
    return res


print("Total:", total(3, 4))
print("Total:", total(3, 8, 12, -5))
```

Listing 5.18 parameters_variable.py File

The following output is generated:

```
Total: 7
Total: 18
```

The function is called with two or four values. The first value is assigned to parameter a. The remaining values are stored in the summands tuple. The for loop is used to add the values of the tuple to the total. The function must be called with at least one parameter.

5.6.2 Named Parameters

The defined order of the parameters does not have to be adhered to when you call a function if you use *named parameters*.

Our next example shows some variants for calling the volume() function. This function calculates the volume of a cuboid and outputs it. The transferred color is output as well.

```python
def volume(width, length, depth, color):
    print("Values:", width, length, depth, color)
    print("Volume:", width * length * depth, "Color:", color)


volume(4, 6, 2, "Red")
```

```
volume(length=2, color="Yellow", depth=7, width=3)
volume(5, depth=2, length=8, color="Blue")
# volume(3, depth=4, length=5, "Black")
```

Listing 5.19 parameters_named.py File

The following output is generated:

```
Values: 4 6 2 Red
Volume: 48 Color: Red
Values: 3 2 7 Yellow
Volume: 42 Color: Yellow
Values: 5 8 2 Blue
Volume: 80 Color: Blue
```

The first call takes place in the familiar form. Only *positional parameters* are used. In other words, the assignment is made via the position of the parameters.

The second call passes four named parameters. In this case, the order in which they are called is not important.

In the third call, one positional and three named parameters are transferred. Mixed forms are therefore also possible.

However, as soon as the first named parameter appears in the call, all subsequent parameters must also be named. An error would therefore occur in the fourth call.

5.6.3 Optional Parameters

Optional parameters also allow a variable number of parameters. They must have a default value and be at the end of the parameters list. Named parameters can be used as well.

Now, let's change the function for calculating the volume of a cuboid. Only two parameters need to be specified. The other two parameters are optional; the default values are used if necessary.

```
def volume(width, length, depth=1, color="Black"):
    print("Values:", width, length, depth, color)
    print("Volume:", width * length * depth, "Color:", color)


volume(4, 6, 2, "Red")
volume(2, 12, 7)
volume(5, 8)
volume(4, 7, color="Red")
```

Listing 5.20 parameters_optional.py File

This program generates the following output:

```
Values: 4 6 2 Red
Volume: 48 Color: Red
Values: 2 12 7 Black
Volume: 168 Color: Black
Values: 5 8 1 Black
Volume: 40 Color: Black
Values: 4 7 1 Red
Volume: 28 Color: Red
```

The depth and color parameters are optional and are positioned at the end of the parameters list. Now, the four calls take place in the following sequence:

- All four parameters are transferred with the first call.
- In the second call, only the first optional parameter is transferred. The second optional parameter is therefore given the default value.
- In the third call, both optional parameters are not transferred and receive their default value.
- In the fourth call, only the second optional parameter is transferred. As this parameter cannot be assigned positionally, it must be named.

5.6.4 Multiple Return Values

Functions in Python can provide a tuple of return values. In this example, the area and circumference of a circle are calculated in the circle() function and returned as a tuple.

```
import math
def circle(radius):
    area = math.pi * radius * radius
    circumference = 2 * math.pi * radius
    return area, circumference

a, c = circle(3)
print(f"Area: {a:.3f}, Circumference: {c:.3f}")
x = circle(3)
print(f"Area: {x[0]:.3f}, Circumference: {x[1]:.3f}")
```

Listing 5.21 return_tuple.py File

The following output is generated:

```
Area: 28.274, Circumference: 18.850
Area: 28.274, Circumference: 18.850
```

The `return` keyword returns a tuple that can be unpacked. This tuple contains the two results of the function.

A tuple of the appropriate size can be available for reception at the call point, as in the first call. However, a variable can also be available, which then becomes a tuple, as in the second call.

5.6.5   Transferring Copies and References

Sometimes, parameters that are passed to a function are changed within the function, which has different effects:

- When an object of a simple type is transferred, for example, a number or a character string, then a copy of the object is created. Changing the copy has no effect on the original.
- When an object of a non-simple type is transferred, for example, a list, a dictionary, or a set, a reference to the original object is used. A change via the reference also changes the original.

To illustrate this correlation, a total of five parameters are passed to a function in the following example: a number, a string, a list, a dictionary, and a set. Each object is output three times:

- Prior to the function call
- After a change within the function
- After returning from the function

```python
def change(za, tx, li, di, st):
    za = 8
    tx = "ciao "
    li[0] = 7
    di["x"] = 7
    st.discard(3)
    print(f"Function: {za} {tx} {li} {di} {st}")

hza = 3s
htx = "hello"
hli = [3,"abc"]
hdi = {"x":3, "y":"abc"}
hst = set([3, "abc"])

print(f"Before:   {hza} {htx} {hli} {hdi} {hst}")
change(hza, htx, hli, hdi, hst)
print(f"After:    {hza} {htx} {hli} {hdi} {hst}")
```

Listing 5.22 parameters_transfer.py File

The output reads:

```
Before:    3 hello [3, 'abc'] {'x': 3, 'y': 'abc'} {3, 'abc'}
Function: 8 ciao  [7, 'abc'] {'x': 7, 'y': 'abc'} {'abc'}
After:     3 hello [7, 'abc'] {'x': 7, 'y': 'abc'} {'abc'}
```

Notice how only the list, the dictionary, and the set were permanently changed by the function. Depending on the problem you want to solve, this effect is either desirable or undesirable.

In the next program, the change of some simple objects is also made permanent by a function. This action makes use of the fact that Python functions can have more than one return value.

The function in this next program sorts two variables in ascending order, which are returned as a tuple; see Chapter 4, Section 4.4.

```python
def sort(one, two):
    if one > two:
        return two, one
    else:
        return one, two


x = 12
y = 7
print(f"x: {x}, y: {y}")
x, y = sort(x, y)
print(f"x: {x}, y: {y}")
```

Listing 5.23 values_change.py File

The following output is generated:

```
x: 12, y: 7
x: 7, y: 12
```

Two numerical values are transferred to the function. The system compares the two values:

- If the first value is greater than the second value, both values are returned to the calling location in reverse order.
- If the reverse is true, the two values are returned to the calling location in the same order.

5.6.6 Namespaces

The definition of a function in Python creates a local namespace. This local namespace contains all the names of the variables to which a value is assigned within the function and the names of the variables from the parameters list.

If a variable is accessed when the function is executed, this variable is first searched for in the local namespace. If the name of the variable is not found there, a search is carried out in the global namespace, i.e., in the previously executed program lines outside the function. If the variable is not found there either, an error will occur. Here's a first example:

```python
def func():
    try:
        print(x)
    except:
        print("Error")


# Program
func()
x = 42
func()
```

Listing 5.24 global_without.py File

The following output is generated:

```
Error
42
```

The first call of `func()` triggers an error because the value of the x variable is supposed to be output in the function. This variable does not exist in the local namespace, nor does it exist in the previously processed program lines outside the function.

The second call of `func()` does not cause an error because the x variable has previously been assigned a value outside the function and is therefore known in the global namespace.

You can use the `global` keyword to assign a variable directly to the global namespace, which is necessary in the next example.

```python
def entry():
    global x
    x = float(input("Number: "))

def output():
    print("Number:", x)
```

```
# Program
entry()
output()
```

Listing 5.25 global_with.py File

In the `entry()` function, a value for the x variable is imported and converted into a number. The x variable is assigned directly to the global namespace using the `global` keyword. Otherwise, the x variable would only be known in the local namespace and would therefore not be found in the `output()` function in either the local or global namespace. The assignment to the global namespace must be made before the x variable is used.

You can test the behavior by temporarily commenting out the line containing the `global` keyword.

5.6.7 Recursive Functions

Certain processes are best programmed recursively. A recursive function calls itself again and again. To prevent this behavior from leading to an endless number of function calls, you must specify a condition that puts an end to the recursion. In addition, an initial call is required to initiate the recursion.

The principle of recursion can be illustrated by a simple program that uses the recursive `halve()` function to halves the value transferred to it each time the function is called. The function then calls itself again with the halved value. The recursion ends after the value, which is constantly halved, has fallen below a certain limit. The first call of the recursive function is made with a start value from the main program.

```
def halve(value):
    print(value)
    value = value / 2
    if value > 0.05:
        halve(value)


# Program
halve(3)
```

Listing 5.26 recursion.py File

The following output is generated:

```
3
1.5
0.75
```

```
0.375
0.1875
0.09375
```

The recursive call is made in compliance with the value > 0.05 condition. Without this condition, the recursion would continue endlessly.

5.6.8   Lambda Functions

Lambda functions are created using the lambda keyword. Also referred to as *anonymous functions*, lambda functions stand in contrast to all the functions used so far in this book, which are referred to as *named functions*.

A lambda function returns its result using an expression that must be in the same line. It must not contain multiple statements, outputs, or loops.

Lambda functions are governed by the following rules:

- Lambda functions have shorter definitions compared to named functions
- They can have parameters
- They can be assigned to a variable through which they can be called
- They can be passed as parameters when another function is called (Section 5.6.9)
- They are structured according to the following pattern: [result] = lambda [parameters list]:[expression]

Let's consider a simple example of a lambda function.

```
multiply = lambda x,y: x*y
add = lambda x,y: x+y

print(multiply(5,3))
print(add(4,7))
```

Listing 5.27 function_lambda.py File

The program generates the following output:

```
15
11
```

In this case, both lambda functions have two parameters. The first lambda function is assigned to the multiply variable and returns the product of the two parameters. The second lambda function is assigned to the add variable and returns the sum of the two parameters.

5.6.9 Functions as Parameters

Not only can you pass values to a function, but you can also pass the name of a named function or an anonymous function. Such a function, which is used as a parameter, is referred to as a *callback* and makes the first function more flexible. Let's look at an example.

```python
def to_the_power_of_two(x):
    return x * x

def to_the_power_of_three(x):
    return x * x * x

def output(bottom, top, step, f):
    for x in range(bottom, top, step):
        print(x, ":", f(x), sep="", end=" ")
    print()

# Program
output(1, 5, 1, to_the_power_of_two)
output(1, 5, 1, to_the_power_of_three)
output(1, 5, 1, lambda x: x * x * x * x)
```

Listing 5.28 parameters_function.py File

The following output is generated:

```
1:1 2:4 3:9 4:16
1:1 2:8 3:27 4:64
1:1 2:16 3:81 4:256
```

First, the named functions to_the_power_of_two() and to_the_power_of_three() are defined. They return a single value as the result.

The output() function is used to output multiple values and the associated function values. It expects a total of four parameters. The first three parameters are used to control the for loop.

The name of a named function (without parentheses) or an anonymous function is expected as the fourth parameter. This function is used to calculate the function value.

5.7 Built-In Functions

Even without integrating an additional module, you have access to a whole range of built-in functions. Table 5.1 contains an overview of the built-in functions covered in this book.

Name	Returns ...	Example in ...
abs()	Amount of a number	Chapter 4, Section 4.1.10
bin()	Binary (dual) number	Chapter 4, Section 4.1.9
bytes()	Object of type bytes	Chapter 4, Section 4.2.9
chr()	Character for Unicode number	Section 5.7.2
filter()	Iterable with filtered values	Section 5.3.3
float()	Number with decimal places	Chapter 3, Section 3.2.4
frozenset()	Non-changeable set	Chapter 4, Section 4.6.1
hex()	Hexadecimal number	Chapter 4, Section 4.1.1
input()	Entered string	Chapter 3, Section 3.2.3
int()	Integer	Chapter 3, Section 3.2.4
len()	Number of elements	Chapter 4, Section 4.2.1
map()	Iterable with results of multiple calls	Section 5.3.2
max()	Largest element	Section 5.7.1
min()	Smallest element	Section 5.7.1
oct()	Octal number	Chapter 4, Section 4.1.1
open()	Reference to an open file	Chapter 8, Section 8.1
ord()	Unicode number for character	Section 5.7.2
print()	Output	Section 5.2.1
range()	Iterable over range	Chapter 3, Section 3.4.6
reversed()	List in reverse order	Section 5.7.3
round()	Rounded number	Chapter 4, Section 4.1.4
set()	Set	Chapter 4, Section 4.6
slice()	Range of a sequence	Chapter 4, Section 4.2.3
sorted()	Sorted list	Section 5.7.3
str()	Character string	Chapter 4, Section 4.2.5
sum()	Sum of the elements	Section 5.7.1

Table 5.1 Some Built-In Functions

Name	Returns ...	Example in ...
type()	Type of an object	Chapter 3, Section 3.2.8
zip()	Connection of iterables	Section 5.3.1

Table 5.1 Some Built-In Functions (Cont.)

5.7.1   The max(), min(), and sum() Functions

The built-in max() and min() functions return the largest or smallest value of an iterable. The sum() function returns the sum of the elements of an iterable and is called with only one parameter.

Let's look at an example using all three functions.

```
print("Max. value of a tuple:", max(3, 2, -7))
print("Min. value of a set:", min(set([3, 2, 3])))
print("Sum of a tuple:", sum((3, 2, -7)))
```

Listing 5.29 max_min_sum.py File

The following output is generated:

```
Max. value of a tuple: 3
Min. value of a set: 2
Sum of a tuple: -2
```

In this case, the max() function is called with several parameters that form a tuple of values, the largest element of which is determined. The min() function is called for a set. The individual parameter of the sum() function is a tuple that must be created before the function call using parentheses.

5.7.2   The chr() and ord() Functions

The built-in chr() function returns the corresponding character for a Unicode number. Conversely, you can use the built-in ord() function to obtain the Unicode number for a character. Let's consider an example.

```
for i in range(48,58):
    print(chr(i), end="")
print()

for i in range(65,91):
    print(chr(i), end="")
print()
```

```
for i in range(97,123):
    print(chr(i), end="")
print()

for z in "abcde":
    print(ord(z), end=" ")
print()

for z in "abcde":
    print(chr(ord(z)+1), end="")
```

Listing 5.30 chr_ord.py File

The following output is generated:

```
0123456789
ABCDEFGHIJKLMNOPQRSTUVWXYZ
abcdefghijklmnopqrstuvwxyz
97 98 99 100 101
bcdef
```

Positions 48 to 57 in Unicode contain the digits 0 to 9. The uppercase and lowercase letters are in positions 65 to 90 and 97 to 122 in Unicode.

In the last part of this program, each character of a string is converted into the character following it in the code. This part could be an example of a simple encryption process.

5.7.3 The reversed() and sorted() Functions

The built-in reversed() function returns the elements of a sequence in reverse order. The built-in sorted() function is used to create and deliver a sorted list of the elements in a sequence.

In our next example, these two functions are applied to a tuple and the values view of a dictionary.

```
t = 4, 12, 6, -2
print("Tuple:", t)
print("Reversed: ", end="")
for i in reversed(t):
    print(i, end=" ")
print()
print("Sorted:", sorted(t))
print()
```

```
dc = {"Peter":31, "Judith":28, "William":35}
print("Dictionary:", dc)
va = dc.values()
print("Values view:", va)
print("Reversed: ", end="")
for i in reversed(va):
    print(i, end=" ")
print()
print("Sorted:", sorted(va))
```

Listing 5.31 reversed_sorted.py File

The following output is generated:

```
Tuple: (4, 12, 6, -2)
Reversed: -2 6 12 4
Sorted: [-2, 4, 6, 12]

Dictionary: {'Peter': 31, 'Judith': 28, 'William': 35}
Values view: dict_values([31, 28, 35])
Reversed: 35 28 31
Sorted: [28, 31, 35]
```

The reversed() function returns a sequence that contains the elements in reverse order. They can be output using a for loop. In the case of the reversed tuple, it is an object of the reversed type; in the case of the dictionary's values view, it is an object of the dict_reversevalueiterator type.

The sorted() function returns a list containing the elements of the sequence in sorted order. Numbers are sorted in ascending order by value; characters are sorted in ascending order by code number.

5.8   Other Mathematical Modules

The mathematically oriented modules in this section represent specific topics. You can skip them for now and look them up later, if necessary.

5.8.1   Drawing Function Graphs

You can use the matplotlib library to create mathematical representations, for example, to draw the graph of a function.

First, the additional matplotlib module must be installed (see also Appendix A, Section A.1) using these commands:

- On Windows and macOS: `pip install matplotlib`
- On Ubuntu Linux: `sudo apt install python3-matplotlib`

In the next program, I will show you numerous possibilities opened up by this library. The two mathematical functions $y1(x) = x^2$ and $y2(x) = 0.5x^2$ are drawn in the range from $x = -5$ to $x = 5$, as shown in Figure 5.12.

```python
import matplotlib.pyplot as plt

x = []
y1 = []
y2 = []
i = -5
while i < 5.05:
    x.append(i)
    y1.append(i * i)
    y2.append(0.5 * i * i)
    i += 0.1

plt.plot(x, y1, label="y1(x) = x * x")
plt.plot(x, y2, label="y2(x) = 0.5 * x * x")
plt.title("Functions")
plt.legend()
plt.xlabel("x")
plt.ylabel("y(x)")
plt.axis([-5, 5, 0, 25])
plt.grid(axis="both")
plt.axhline(9, color="red")
plt.axvline(3, color="green")
plt.show()
```

Listing 5.32 matplotlib_graph.py File

First, the `pyplot` module is imported from the `matplotlib` module. The `plt` alias is created for the calls in the program using the as keyword. The module can be accessed via this keyword in abbreviated form.

Three empty lists are created, which are filled in a loop: the x list with the x-values from -5 to 5 at an increment of 0.1 and the y1 and y2 lists with the respective function values.

Each time the `plot()` function is called, a single graph is generated for the values from two lists. In our example, the method is called twice: once for the values from the x and y1 lists and once for the values from the x and y2 lists. The optional `label` parameter can be used together with the `legend()` function to create a legend for your plot. The `show()` function draws the generated graphs in an application window on the screen.

Additional functions are available when designing the drawing, such as the following:

- `title()` sets a title above the drawing.
- `legend()` displays a legend, together with the named `label` parameter of the `plot()` function.
- `xlabel()` and `ylabel()` are used for labeling the two axes.
- `axis()` set the limits of the two axes using a list of four values.
- `grid()` displays a grid parallel to the axes; possible values for the named `axis` parameter are x, y, or, both (the default value).
- `axhline()` displays a horizontal line at a specific y-value in a specific color.
- `axvline()` displays a vertical line at a specific x-value in a specific color.

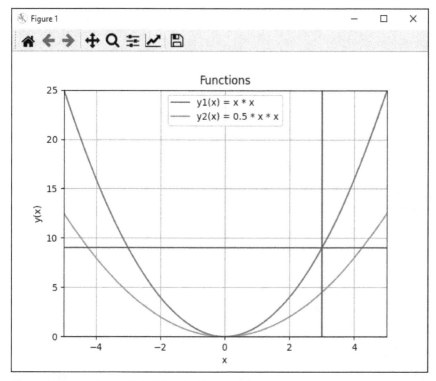

Figure 5.12 Functions y1(x) = ^{x2} and y2(x) = 0.5x^2

5.8.2   Multiple Partial Drawings

Another program shows how different graphs can be displayed in several partial drawings using the `matplotlib` library. Our example has two functions, y1(x) = sin(x) and y2(x) = cos(x), in the range from 0 to 360 degrees, as shown in Figure 5.13.

```python
import matplotlib.pyplot as plt
import math

x = []
y1 = []
y2 = []
i = 0
for i in range(0, 361):
    x.append(i)
    bm = math.radians(i)
    y1.append(math.sin(bm))
    y2.append(math.cos(bm))

drawing, (py1, py2) = plt.subplots(2, 1)

py1.plot(x, y1)
py1.set_title("Sine")
py1.set_xlabel("x")
py1.set_ylabel("sin(x)")
py1.axis([0, 360, -1, 1])
py1.grid(axis="both")

py2.plot(x, y2)
py2.set_title("Cosine")
py2.set_xlabel("x")
py2.set_ylabel("cos(x)")
py2.axis([0, 360, -1, 1])
py2.grid(axis="both")

plt.show()
```

Listing 5.33 matplotlib_subplot.py File

A total of three empty lists are created, which are filled in a loop: The x list with the x-values from 0 to 360 at an increment of 1 and the y1 and y2 lists with the corresponding function values, which are determined using the math.sin() and math.cos() functions. The degree values are first converted into radians using the math.radians() function.

The subplots() function is used to subdivide a drawing into multiple subplots. These are arranged in a grid of rows and columns using the parameters. Here there are two rows and one column. So, there are two partial drawings. A reference to the entire drawing and a suitable number (here: 2) of references to the partial drawings are returned.

Multiple functions are called for each of the partial drawings:

- set_title() sets a title above the partial drawing.
- set_xlabel() and set_ylabel() are used to label the two axes.
- axis() sets the limit values for the two axes.
- grid() creates a grid parallel to the axes.

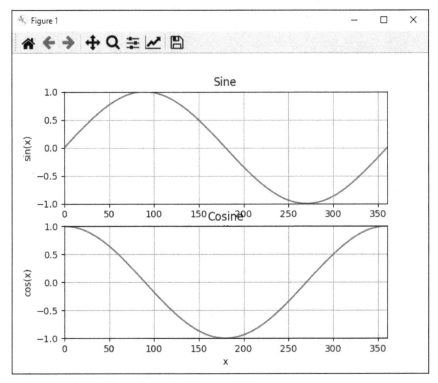

Figure 5.13 The y1(x) = sin(x) and y2(x) = cos(x) Functions

5.8.3 One-Dimensional Arrays and Vectors

The numpy library provides numerical functions, which are required for mathematical approximations, for example. These functions also make it easier to work with large, multidimensional data structures, such as those used in calculations with vectors and matrices.

After installing the matplotlib module, the numpy module is already installed, as it is required internally. Otherwise, you can install it with the following commands (see also Appendix A, Section A.1):

- On Windows and macOS: pip install numpy
- On Ubuntu Linux: sudo apt install python3-numpy

In our first example, the program works with one-dimensional data structures to solve some *vector calculation* tasks.

```python
import numpy as np

a = np.array([5, 8, -4])
print("One-dimensional array = vector:", a)

b = a * 2
print("Arithmetic operation with scalar:", b)

c = np.linspace(0, 30, 4)
print("Array with equidistant values:", c)

d = np.sin(np.radians(c))
print("Apply function to array:", d)

e = np.array([2, 8, 3])
f = np.array([3, 7, -4])
g = e + f
print("Vector addition:", g)

h = e @ f
print("Scalar product:", h)

i = np.cross(e, f)
print("Cross product:", i)
```

Listing 5.34 numpy_one_dim.py File

The numpy module is imported and made accessible via the np alias according to convention. The array() function is used to create a one-dimensional array from a list. The array has three elements and can also be regarded as a vector in three-dimensional space.

Calculation operations with scalars are carried out for the entire array. The multiplication of the vector with the value 2 serves as our example. As in vector calculus, these operations are applied element by element. The mathematical foundations for understanding vector calculus are not the subject of this book, however. We recommend referring to *https://en.wikipedia.org/wiki/Euclidean_vector*.

The linspace() function creates an array containing equidistant values, such as those required for the x-axis of a drawing with matplotlib. The first two parameters define the first and last value of the array; the third parameter defines the number of values in the array.

The numpy module provides trigonometric functions that work more effectively and accurately than the corresponding functions from the math module. These functions from the numpy module are also executed for the entire array, element by element. The radians() and sin() functions are used in our example to convert to radians and calculate the sine, respectively.

Calculations with vectors are carried out according to the rules of vector calculus for vectors of the same size:

- *Vector addition*, using the + operator
- *Scalar product*, using the @ operator
- *Cross product*, using the cross() function

Here's the output of the program:

```
One-dimensional array = vector: [ 5  8 -4]
Arithmetic operation with scalar: [10 16 -8]
Array with equidistant values: [ 0. 10. 20. 30.]
Apply function to array: [0.         0.17364818 0.34202014 0.5        ]
Vector addition: [ 5 15 -1]
Scalar product: 50
Cross product: [-53  17 -10]
```

5.8.4 Multidimensional Arrays and Matrices

This section deals with multidimensional arrays and matrices that are created and edited using the numpy library. Let's start by distinguishing between arrays and matrices:

- Arrays can have any number of dimensions. A three-dimensional array is generated in our example.
- Matrices can only be two-dimensional. In our example, they are used to solve some problems from *matrix calculation*.

Let's take a look at the program first.

```python
import numpy as np

a = np.array([[[2, 8],[3, 7]], [[5, 4],[9, 1]]])
b = a * 2
print("Apply operator and scalar to multidimensional array:")
print(b)
print("Shape of the array:", b.shape)
print()

c = np.matrix([[5, 2, 3],[8, 4, 5]])
```

```
print("Matrix c:")
print(c)
print("Shape of the matrix:", c.shape)
print()

d = c * 2
print("Arithmetic operation with scalar:")
print(d)
print()

e = np.matrix([[2, -4, 7],[-1, 9, 3]])
print("Matrix e:")
print(e)
print("Shape of the matrix:", e.shape)
print()

f = c + e
print("Matrix addition c + e:")
print(f)
print()

g = np.matrix([[2, 3],[4, 5],[2, 6]])
print("Matrix g:")
print(g)
print("Shape of the matrix:", g.shape)
print()

h = c * g
print("Matrix multiplication c * g:")
print(h)
```

Listing 5.35 numpy_multi_dim.py File

First, the array() function is used to create a three-dimensional array from a three-dimensional list. Only two dimensions can be clearly displayed on the screen. Blank lines are therefore automatically inserted to clarify the next higher dimension.

The entire three-dimensional array is used to perform an arithmetic operation with a scalar, again element by element.

The shape property returns a tuple with the size of an array in the individual dimensions. This array has 2 × 2 × 2 elements.

The next step is to create the c matrix with two rows and three columns using the matrix() function. This matrix also has the shape property.

Calculation operations with scalars are carried out for the entire matrix. As with matrix calculation, these operations are also applied element by element. The mathematical foundations for understanding matrix calculation are not the subject of this book, however. We recommend referring to *https://en.wikipedia.org/wiki/Matrix_(mathematics)*.

The e matrix is generated. With two rows and three columns, this matrix has the same shape as the c matrix. For this reason, these two matrixes can be added based on the rules of *matrix addition*.

The g matrix is then created with three rows and two columns. It therefore has the right shape to be multiplied by matrix c based on the rules of *matrix multiplication*.

The following output is generated:

```
Apply operator and scalar to multidimensional array:
[[[ 4 16]
  [ 6 14]]

 [[10  8]
  [18  2]]]
Shape of the array: (2, 2, 2)

Matrix c:
[[5 2 3]
 [8 4 5]]
Shape of the matrix: (2, 3)

Arithmetic operation with scalar:
[[10  4  6]
 [16  8 10]]

Matrix e:
[[ 2 -4  7]
 [-1  9  3]]
Shape of the matrix: (2, 3)

Matrix addition c + e:
[[ 7 -2 10]
 [ 7 13  8]]

Matrix g:
[[2 3]
 [4 5]
 [2 6]]
Shape of the matrix: (3, 2)
```

```
Matrix multiplication c * g:
[[24 43]
 [42 74]]
```

5.8.5 Signal Processing

The scipy library provides numerous functions for scientific calculations. Based on the numpy library, scipy contains, among other things, the io module for reading and writing values from files. The io module in turn contains the wavfile module with the read() and write() functions for reading and writing digital signals from WAV files.

You can install the scipy module with the following commands (see also Appendix A, Section A.1):

- On Windows and macOS: pip install scipy
- On Ubuntu Linux: sudo apt install python3-scipy

In our next program, digital audio signals are read from a WAV file and displayed separately for the left and right channels in a drawing, as shown in Figure 5.14. In addition, some information about the WAV file is output.

```python
import scipy.io.wavfile as wf
import matplotlib.pyplot as plt
import numpy as np

samplingrate, values = wf.read("fanfare.wav")
print(f"Sampling rate: {samplingrate} Values / sec.")
print("Number of values:", values.shape[0])
length = values.shape[0] / samplingrate
print(f"Length: {round(length,2)} sec.")

time = np.linspace(0, length, values.shape[0])
plt.plot(time, values[:, 0], label="Left")
plt.plot(time, values[:, 1], label="Right")
plt.legend()
plt.xlabel("sec.")
plt.ylabel("Amplitude")
plt.show()
```

Listing 5.36 scipy_wav.py File

The additional output reads as follows:

```
Sampling rate: 44100 Values / sec.
Number of values: 71296
Length: 1.62 sec.
```

182

The read() function returns a tuple with two elements:

- The first element contains the sampling rate of the WAV file. This rate corresponds to the frequency at which the original analog audio signal was digitally sampled. It is measured in the "Hz" unit or "values per second". A common sampling rate is 44.1 KHz (i.e., 44,100 values per second).

- The second element contains a two-dimensional numpy array with the stored values from the WAV file for the amplitudes of the two channels.

The number of saved values can be determined using the shape property. The length of the audio signal (in seconds) can be calculated from the sampling rate and the number of values.

The linspace() function is used to create a list of values for the x-axis of the drawing. The length of the audio signal over time is plotted on the x-axis. The plot() function is called for each channel of the audio signal. A title, a legend, and the axis labels are also created.

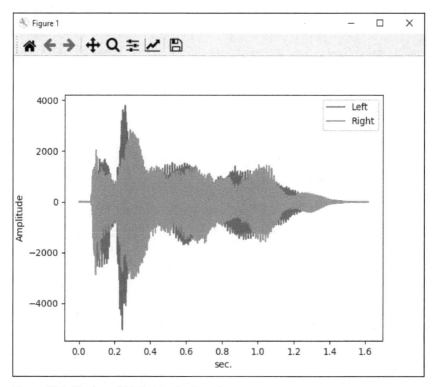

Figure 5.14 Display of Digital Audio Signals

5.8.6   Statistics Functions

With statistical methods, you can analyze large quantities of values and determine certain representative information about these values. These values may derive from surveys or other samples, for example. Since version 3.4, Python provides some useful functions for iterables from these values in the statistics module. Python 3.6 and Python 3.8 have added a few more functions.

Let's look at our example program.

```python
import statistics as sta

pa = [5, 2, 4, 17]
print("List:", pa)
print("Arithmetic mean:", sta.mean(pa))
print("Geometric mean:", sta.geometric_mean(pa))
print("Harmonic mean:", sta.harmonic_mean(pa))
print("Median:", sta.median(pa))
print("Lower median:", sta.median_low(pa))
print("Upper median:", sta.median_high(pa))
print()

pb = [5, 2, 4, 17, 3]
print("List:", pb)
print("Median:", sta.median(pb))
print("Lower median:", sta.median_low(pb))
print("Upper median:", sta.median_high(pb))
print()

pc = [3, 5, 5, 12, 17, 17]
print("Mode:", sta.mode(pc))
print("Multimode:", sta.multimode(pc))
print()

print("Arithm. mean from tuple:", sta.mean((5, 2, 4, 17)))
pd = {'NY':5, 'TX':2, 'CA':4, 'LA':17}
print("Arithm. mean from dictionary:", sta.mean(pe.values()))
```

Listing 5.37 statistic_functions.py File

The following output is generated:

```
List: [5, 2, 4, 17]
Arithmetic mean: 7
Geometric mean: 5.1065457621381
Harmonic mean: 3.9650145772594754
Median: 4.5
Lower median: 4
Upper median: 5
```

```
List: [5, 2, 4, 17, 3]
Median: 4
Lower median: 4
Upper median: 4

Mode: 5
Multimode: [5, 17]

Arithm. mean from tuple: 7
Arithm. mean from dictionary: 7
```

First, the statistics module is imported. For the calls in the program, this module can be accessed as sta for short using the as keyword.

The mean() function returns the arithmetic mean of a series of values, that is, the sum of the values divided by their number. In our example, this calculation corresponds to the following operations: (5 + 2 + 4 + 17) / 4.

The geometric_mean() function has been available since Python 3.8. This function provides the geometric mean of a series of values, which corresponds to the n-th root of the product of n values. In this case, this calculation would be the fourth root of (5 × 2 × 4 × 17).

The harmonic_mean() function has been available since Python 3.6. This function provides the harmonic mean value, which corresponds to the number of values divided by the sum of their reciprocal values. In our example, this would be 4 / (1/5 + 1/2 + 1/4 + 1/17).

Further usage examples for various mean values can be found at *https://en.wikipedia.org/wiki/Mean*.

The median() function returns the median. Also called the "central value", this value is at the center of the number set: There are just as many values that are greater than the median as there are values that are less than the median. In the case of an odd set of values, the median is the element in the middle of the set of numbers. In the case of an even set of values, the median is the arithmetic mean of the two elements in the middle of the number set.

The median_low() and median_high() functions return the lower median and the upper median, respectively. These elements always exist in a number set. For an odd set of values, it is for both functions the element in the middle of the number set. For an even set of values, these are the two elements in the middle of the number set.

The mode() function returns the value that occurs most frequently in the number set. If multiple values have the highest frequency, the value that occurs first is named. The multimode() function, available since Python 3.8, provides a list of all values with the highest frequency.

The series of numbers examined can also derive from a tuple or, using the values() method, from the values view of a dictionary.

5.9 Custom Modules

We've already used quite a few functions from standard modules such as random, math, fractions, copy, and statistics as well as from installed modules such as matplotlib, numpy, and scipy.

However, you can also define and import your own modules. This capability is useful if certain custom functions (or classes, see Chapter 6) are required by different programs. The creation and use of your own modules is quite easy in Python and is the subject of this section.

5.9.1 Creating Custom Modules

To create a module, you simply need to save the desired function (or class) in a separate file. The name of the file is also the name of the module. You can create the function as usual.

```
def square(x):
    return x * x
```

Listing 5.38 module_new.py File

You can then use this function in any program once you have imported it. Several options exist for importing a module, which I will describe next.

5.9.2 Standard Import of a Module

Our first example illustrates the usual way to import a module.

```
import module_new
print("Square:", module_new.square(3))
```

Listing 5.39 module_use.py File

All functions (or classes) of the module_new module (i.e., the *module_new.py* file) are made accessible by using the import statement. The square() function can be called with the following command structure:

```
[module_name].[function_name]
```

5.9.3 Importing and Renaming a Module

If a standard module or your own module has a long, unwieldy name, you can give the module an alias using as, as shown in our next example.

```
import module_new as mnew
print("Square:", mnew.square(3))
```

Listing 5.40 module_as.py File

In this case, you can access the module_new module by using mnew.

5.9.4 Importing Functions

The from keyword enables you to import specific functions or all functions (or classes) from a module.

```
from module_new import square
print("Square:", square(3))
```

Listing 5.41 module_from.py File

The square() function is imported from the module_new module via the from statement. You can then call it without the module name, as if it were defined in the same file. The other functions of the module are not available.

In contrast, you can also import all functions from the module_new module at once. The statement then reads:

```
from module_new import *
```

I am only mentioning the import using the from keyword for the sake of completeness. This option is no longer recommended and should only be used outside of functions. The from keyword is also not useful if you want to import multiple functions with the same name from different modules.

> **Note**
> In our examples, we assume that modules are located in the same directories as the programs that use them.

5.9.5 Exercise

> **u_module**
> Rewrite the program from the u_return_value exercise from Chapter 3, Section 3.7.7: Now, we want to swap out the tax() function to the *u_module_finance.py* file. We

187

want the main program in the *u_module.py* file to import and use the function from that file.

5.10   Command Line Parameters

As mentioned earlier, a Python program can be called from the command line of the operating system. In Chapter 2, Section 2.3.2 and in Section 2.3.3, I described how you can access the command line and the terminal.

The call is similar to calling a function with parameters. The individual parameters are separated from each other by spaces. These parameters are available within the program in the argv list from the sys module. The number of these parameters can be determined using the built-in len() function.

Let's look at a program that can be called with any number of parameters. This program determines the total of those parameters that contain valid numerical values.

```
import sys
print("Parameter:", sys.argv)
print("Number:", len(sys.argv))

total = 0
for p in sys.argv[1:]:
    try:
        total += float(p)
    except:
        print(f"Error at parameter {p}")
print("Total:",total)
```

Listing 5.42 command.py File

This program is called with the following command line:

```
python command.py 12 abc -3.5
```

The following output is generated:

```
Parameter: ['command.py', '12', 'abc', '-3.5']
Number: 4
Error at parameter abc
Total: 8.5
```

First, the list of parameters is displayed together with their numbers. The first element of the sys.argv list is the name of the program. The other elements represent the parameters to be used. They can be run through using the sys.argv[1:] slice and a for loop. If one of the usage parameters cannot be converted, an error message will display.

5.11   "Fractions Training" Program

In this section, we'll write a training program for doing fractions with three levels of difficulty. Both the numerator and the denominator of the occurring fractions can be positive or negative.

This larger example can be skimmed for the time being. We simply wanted to illustrate the interaction of many familiar elements, such as multiple branches, loops, exception handling, tuples, and functions with multiple return values.

5.11.1   The Program Flow

To illustrate how this program should flow, I will first explain how the program works by using a few sample entries:

```
Your selection: 1 = Easy, 2 = Medium, 3 = Difficult, 0 = End
1
Calculate integer: 48 / -8
Result: -6
Your selection: 1 = Easy, 2 = Medium, 3 = Difficult, 0 = End
2
Reduce fraction: -42 / -12
Result: 7 / 2
Your selection: 1 = Easy, 2 = Medium, 3 = Difficult, 0 = End
3
Calculate result fraction: -10/6 + 10/-6
Result: -10 / 3
Your selection: 1 = Easy, 2 = Medium, 3 = Difficult, 0 = End
0
```

After starting the program, a menu appears with which an easy task (input "1"), a moderately difficult task (input "2"), or a difficult task (input "3") can be selected. After editing a task, the menu appears again. Here you can either select another task or end the program (enter "0").

After selecting an easy task, a fraction is displayed, which results in an integer. The fraction line is represented by a division line between the numerator and denominator. You should now calculate the integer that represents the result of the task either mentally or on paper. After pressing Enter , the correct result appears, which you can

189

compare with your own result. After correctly solving (for example) 20 easy tasks, a moderately difficult task can be selected next.

After selecting a moderately difficult task, a fraction is displayed that is to be reduced to the smallest possible fraction. The fraction that represents the result of the task should now be calculated mentally or on paper. After pressing [Enter], the correct result appears again, which you can compare with your own result. After correctly solving (for example) 20 moderately difficult tasks, a difficult task can be selected next.

After selecting a difficult task, two fractions are displayed that are to be added, subtracted, divided, or multiplied. The next step is to reduce the result of the calculation to the smallest possible fraction. After pressing [Enter], the correct result will also appear, which you can compare with your own result.

5.11.2 Main Program

I will explain the program in individual parts. First comes the main program, which is at the end of the file.

```
# Program
while True:
    while True:
        print("Your selection: 1 = Easy, 2 = Medium, 3 = Difficult, 0 = End")
        try:
            selection = int(input())
            if selection < 0 or selection > 3:
                raise
            else:
                break
        except:
            print("Please enter only 0, 1, 2, or 3")

    if selection == 0:
        break
    elif selection == 1:
        easy()
    elif selection == 2:
        medium()
    else:
        difficult()
```

Listing 5.43 fraction_training.py File, Main Program

The menu and the input are embedded in an external infinite loop that can only be exited by entering "0".

The menu display is embedded in an internal infinite loop. If an entry is made that cannot be converted into an integer, an exception occurs. If a number is entered that does not correspond to a value in the menu, a separate exception is created using the raise keyword. After an exception, an error message appears, and the menu is displayed again.

If a valid number is entered, one of the three functions easy(), medium(), or hard() is called, or the outer endless loop is exited.

5.11.3 An Easy Task

Let's look at the easy() function to create an easy task. The entire program also requires an initialized random generator.

```
import random
random.seed
...
def easy():
    factor = random.randint(-10, 10)
    d = 0
    while d == 0:
        d = random.randint(-10, 10)
    n = factor * d
    print(f"Calculate integer: {n}/{d}")
    input()
    print("Result:", factor)
```

Listing 5.44 fraction_training.py File, easy() Function

The random generator returns integers for a factor and for the denominator of the fraction. The numerator of the fraction corresponds to the denominator, multiplied by the factor. For example, let's say that the factor 3 and the denominator 5 result in the numerator 15. The task 15 / 5 is therefore determined and output. As a result, (the factor) 3 should be calculated.

Due to the range from -10 to 10, the random generator can return the value 0. The while loop is used to ensure that a new denominator is determined, as the division by 0 is mathematically not permitted.

5.11.4 A Moderately Difficult Task

The medium() function for creating a moderately difficult task follows, which requires the prob tuple and the product() and output() functions.

```
...
prob = 2, 2, 2, 2, 3, 3, 3, 5, 5, 7
...
def product(number):
    value = 1
    for i in range(number):
        value *= random.choice(prob)
    if random.randint(0,1) == 0:
        return value
    else:
        return -value

def output(n, d):
    gcd = math.gcd(n, d)
    n = int(n / gcd)
    d = int(d / gcd)
    if d < 0:
        n *= -1
        d *= -1
    print("Result:", n if d == 1 else f"{n}/{d}")
...
def medium():
    n = product(3)
    d = product(3)
    print(f"Reduce fraction: {n}/{d}")
    input()
    output(n, d)
```

Listing 5.45 fraction_training.py File, medium() Function

In the medium() function, the n and d variables contain the numerator and denominator of the fraction. They are each generated by calling the product() function.

The product() function creates and returns a value that is generated using the random generator. For this purpose, the for loop is first run through three times. With each run, the choice() function from the random module returns a random value from the prob tuple. It contains the values 2, 3, 5, and 7 with different *probabilities*. This approach ensures that the tasks contain smaller rather than larger numbers. The three values are multiplied with each other.

The sign of the calculated product is then set, again using the random generator. If the random generator returns a 0, the product remains positive. If it returns a 1, the product is multiplied by –1.

The output() function is called to display the result. The numerator and denominator are first reduced by dividing them by their greatest common divisor (GCD).

If the result is a negative fraction, the sign should be in the numerator. If both the numerator and denominator are negative, the fraction should be converted into the corresponding positive fraction. If the denominator is negative, the numerator and denominator are multiplied by –1. Two examples: 4 / –3 becomes –4 / 3, while –4 / –3 becomes 4 / 3.

If the denominator has the value 1, only the numerator will be output.

The result of your own calculation is compared with the reduced and, if necessary, converted fraction.

5.11.5 A Difficult Task

The difficult() function to create a difficult task follows, which also requires the ops tuple.

```
...
ops = '+', '-', '*', '/'
...
def difficult():
    an = product(2)
    ad = product(2)
    bn = product(2)
    bd = product(2)
    op = random.choice(ops)
    print(f"Calculate result fraction: {an}/{ad} {op} {bn}/{bd}")

    if op == "+":
        n = an * bd + ad * bn
        d = ad * bd
    elif op == "-":
        n = an * bd - ad * bn
        d = ad * bd
    elif op == "*":
        n = an * bn
        d = ad * bd
    else:
        n = an * bd
        d = ad * bn

    input()
    output(n, d)
```

Listing 5.46 fraction_training.py File, difficult() Function

The ops tuple contains four characters as elements, namely, the operators for the four different calculations.

The numerator and denominator of the two fractions—let's call them fractions A and B—are generated using the product() function. Fraction A has the numerator an and the denominator ad, while fraction B has the numerator bn and the denominator bd. The four values are each the product of two random numbers.

The operator for the task is also generated randomly. The task is then displayed in the form A <operator> B.

Using multiple branches, the result is determined depending on the operator and in compliance with the rules of fractional arithmetic. In this case too, the n and d variables contain the numerator and denominator of the result. The output() function is called to display the result.

Chapter 6
Object-Oriented Programming

This chapter provides an introduction to *object-oriented programming (OOP)*. I'll explain how creating a class hierarchy can help you work on large software projects using classes, objects, properties, and methods and following the principle of inheritance.

6.1 Basic Principles

OOP provides additional options for improving the structure of programs and simplifying their maintenance and expansion. These advantages are particularly evident in large programming projects.

6.1.1 What Is Object-Oriented Programming?

In OOP, classes are created in which the properties and methods of objects are defined. Methods are functions that can only be applied to objects of the relevant class. The properties and methods together form the *members* of a class.

You can create different objects of these classes, assign different values to the properties, and apply methods to these objects. Methods are often used to change the values of an object's properties. The definitions from the class and the assigned values accompany an object throughout its entire lifecycle for the duration of the program.

Let's look at an example: A custom Vehicle class is created in which the properties and methods of vehicles are defined. Among other things, a vehicle has the following properties: name, speed, and direction_of_travel. You can also *accelerate* and *steer* a vehicle. The accelerate() method changes the value of the speed property of an object. Many different vehicles can be created and used within the program.

Classes can inherit their properties and methods. Such a class acts as a *base class*, while its heirs are called *derived classes*. Inheritance thus simplifies the definition of similar objects that have a number of common properties and methods.

Let's consider another example: We create a Car class and a Truck class. Both classes are derived from the Vehicle base class and inherit all the properties and methods of this class. They also have their own properties and methods that are particularly important for the respective class. A Car has a certain number of passengers, and you can get *in* and *out*. A Truck has a load that can be *loaded* and *unloaded*.

6.1.2   Classes, Objects, and Custom Methods

As an example, we'll define our custom class Vehicle. At first, an object of this class only has the speed property and the accelerate() and output() methods. The output() method outputs the current status of the Vehicle object.

Let's look more closely at the definition of the Vehicle class.

```
class Vehicle:
    speed = 0
    def accelerate(self, value):
        self.speed += value
    def output(self):
        print("Speed:", self.speed)
```

Listing 6.1 oop_class.py File, Class Definition

The definition of the class is introduced by the class keyword, followed by the name of the class and a colon. By convention, the name of a class should begin with a capital letter. The other lines of the definition are indented. The speed property is defined, and its value is set to 0.

Methods are defined like functions, using the def keyword. They have at least one parameter, namely, a reference to the object itself. By convention, this reference is called self.

The accelerate() method has a total of two parameters: The first parameter is the reference to the object itself. The second parameter is the value for the change in speed. Within the method, this second parameter is used to change the value of the property of the object. The output() method has only one parameter: the reference to the object itself. This method is simply used to output the properties of an object.

Up to this point, the program only contains a class definition, but the class does not execute anything. Let's look at our main program a bit more closely.

```
volvo = Vehicle()
ford = Vehicle()

volvo.output()
volvo.accelerate(20)
volvo.output()

ford.output()
```

Listing 6.2 oop_class.py File, Main Program

This program generates the following output:

```
Speed: 0
Speed: 20
Speed: 0
```

In the main program, two objects of the Vehicle class are first created; their names are volvo and ford. This process is also referred to as *instantiating* an object; that is, you are creating objects that are *instances of a class*.

When you then start calling the methods, note that the object itself is not specified as a parameter. When a method is called, it always receives one parameter less than specified in the definition. This is because the method is called *for* a specific object. For this reason, within the method, it is known which object it is, usually with the help of self.

The speed of the volvo object is output once before and once after acceleration. The speed of the ford object is only output once. At first (i.e., after their creation), the objects have a speed of 0 as specified in the class definition.

Note

In most object-oriented languages, you can encapsulate members within a class. These members, called *private members*, can only be accessed within the relevant class. This limitation is intended, for example, to prevent the value of a property from being changed inadvertently.

In contrast to these restricted members of a class, *public members* can be reached from outside the class in which they are defined.

In Python, the properties of a class cannot be encapsulated. However, one convention you can follow is to start the name of the property with an underscore (_), for example, _speed = 0 and self._speed += value. In this way, you can document that this property is to be protected.

6.1.3   Special Members

Some special members can often be defined or used in connection with a class. These special members are easily recognizable because their names begin and end in double underscores.

Let's explore a few kinds of special members in a class:

- You can define a special method with the fixed name __init__() as a *constructor method*, which initializes an object with initial values at the beginning of its lifecycle. You can also use optional (see Chapter 5, Section 5.6.3) and named parameters (see Chapter 5, Section 5.6.2) so that objects can receive their initial values in different ways.

- You can define a special method with the specified name __str__(), which outputs the properties of an object.

- The __dict__ property provides a dictionary with the names and values of the properties.

- You can define a special method with the specified name __del__() as a *destructor method*, which triggers actions at the end of an object's lifecycle, such as closing an open file.

Now, we want to change our Vehicle class: In addition to the speed property, a vehicle is assigned the name property, a constructor method, an output method, and a destructor method.

Let's explore this program in its modified form.

```python
class Vehicle:
    def __init__(self, n="(empty)", s=0):
        self.name = n
        self.speed = s
    def __str__(self):
        return f"{self.name} {self.speed} mph"
    def __del__(self):
        print(f"Removed: {self}")
    def accelerate(self, value):
        self.speed += value

ford = Vehicle("Ford Explorer", 40)
renault = Vehicle("Renault Espace")
fiat = Vehicle(s=60)
mercedes = Vehicle()

print(ford)
print(renault)
print(fiat)
```

```
print(mercedes)

ford.accelerate(20)
print(ford.__dict__)
del ford
# print(ford)
```

Listing 6.3 oop_special.py File

6

The output of the program reads:

```
Ford Explorer 40 mph
Renault Espace 0 mph
(empty) 60 mph
(empty) 0 mph
{'name': 'Ford Explorer', 'speed': 60}
Removed: Ford Explorer 60 mph
```

The constructor method has two parameters, which are used to create the two properties and assign initial values to them. Both parameters are optional in this case.

If the second parameter is missing, its default value is used. If no value is available for the first parameter, the second parameter can be passed as a named parameter. If both parameters are missing, both default values will be used.

The output method returns a string with the values of an object's properties.

You can use the __dict__ property for data exchange using JavaScript Object Notation (JSON), for example (see Chapter 8, Section 8.5).

Only brief information is output in the destructor method defined in this example. This method is called using the del keyword. The object then no longer exists. Note also, in the destructor method, the call of the output method using self.

Note

If you have already worked with other programming languages, note that methods and functions cannot be overloaded in Python. If you define a method multiple times, possibly with different parameters, only the most recent definition will apply. However, methods can be overwritten in derived classes (Section 6.2.1).

6.1.4 Exercise

---

u_oop_point

This exercise and some other exercises in this chapter deal with the topic of two-dimensional *geometry*. Develop a class for representing points in a Python file whose

---

contents are to be imported. A point has an x-coordinate and a y-coordinate, which indicate its position in a two-dimensional coordinate system.

The class should enable the following Python program:

```
from u_oop_geometry import *
point1 = Point()
print(point1)
point2 = Point(3.5, 2.5)
print(point2)
point3 = Point(4.0)
print(point3)
point4 = Point(yk=1.5)
print(point4)
point4.move(4.5, 2)
print(point4)
```

Listing 6.4 The u_oop_point.py File

The planned output reads as follows:

```
(0.0 / 0.0)
(3.5 / 2.5)
(4.0 / 0.0)
(0.0 / 1.5)
(4.5 / 3.5)
```

All further information on the solution can be found in the program and the output.

6.1.5   Operator Methods

Operator methods are special methods that are called for objects using operators. These methods are already predefined for built-in data types. For your own data types (i.e., your own classes), you can define operator methods yourself.

In our next program, some calculations with fractions are carried out according to the rules of mathematical fraction calculation (see *https://en.wikipedia.org/wiki/Fraction*). At the end, some comparisons between fractions are made. In contrast to Chapter 4, Section 4.1.10, the methods in this example are defined in our own Fraction class.

Let's start with the main program.

```
...
b1 = Fraction(3, -2)
b2 = Fraction(1, 4)

b3 = b1 * b2
print(f"{b1} * {b2} = {b3}")
```

```
print(f"{b1} / {b2} = {b1 / b2}")
print(f"{b1} + {b2} = {b1 + b2}")
print(f"{b1} - {b2} = {b1 - b2}")
print(f"{b1} + {b2} * {b3}= {b1 + b2 * b3}")
print(f"({b1} + {b2}) * {b3}= {(b1 + b2) * b3}")

b4 = Fraction(-30, 20)
print(f"{b1} == {b4}: {b1 == b4}")
print(f"{b2} > {b4}: {b2 > b4}")
print(f"{b2} < {b4}: {b2 < b4}")
```

Listing 6.5 oop_operator.py File, Main Program

Two objects of our own Fraction class are created. The *, /, +, and - operators are used to perform calculations. Then, you see that the use of parentheses also influences the order of the calculation. Finally, two fractions are compared using the ==, >, and < operators.

Here's the output of the program:

```
-3/2 * 1/4 = -3/8
-3/2 / 1/4 = -6
-3/2 + 1/4 = -5/4
-3/2 - 1/4 = -7/4
-3/2 + 1/4 * -3/8= -51/32
(-3/2 + 1/4) * -3/8= 15/32
-3/2 == -3/2: True
1/4 > -3/2: True
1/4 < -3/2: False
```

Let's look closely at the first part of the definition of our own Fraction class.

```
import math

class Fraction:
    def __init__(self, numerator=1, denominator=1):
        self.n = numerator
        self.d = denominator if denominator != 0 else 1
    def reduce(self):
        g = math.gcd(self.n, self.d)
        self.n = int(self.n / g)
        self.d = int(self.d / g)
        if self.d < 0:
            self.n *= -1
            self.d *= -1
    def value(self):
```

```
        return self.n / self.d
    def __str__(self):
        self.reduce()
        return f"{self.n}/{self.d}" if self.d != 1 else f"{self.n}"
...
```

Listing 6.6 oop_operator.py File, Fraction Class, Part 1 of 2

The numerator and denominator each receive the default value 1 in the constructor. If a fraction with a denominator of 0 is created, the denominator will be set to 1.

In various situations, for example, before the output step, the fraction is reduced. For this purpose, the reduce() method first calculates the greatest common divisor (GCD) of the numerator and denominator using the gcd() function. The numerator and denominator are then divided by that value.

If the denominator is negative, the numerator and denominator will be multiplied by –1. If the fraction has exactly one negative sign in the numerator and denominator, the negative sign will be placed in front of the numerator. If the fraction has two negative signs, both negative signs will disappear.

The value() method calculates the mathematical value of a fraction as a float number. The value is required for comparisons, for example.

The output checks whether the denominator has the value 1. If this is the case, only the numerator will be output. Otherwise, the numerator and denominator are output, separated by a forward slash.

Now, we'll write the operator methods of our own Fraction class.

```
    def __mul__(self, other):
        return Fraction(self.n * other.n, self.d * other.d)
    def __truediv__(self, other):
        return Fraction(self.n * other.d, self.d * other.n)
    def __add__(self, other):
        return Fraction(self.n * other.d + self.d * other.n,
                        self.d * other.d)
    def __sub__(self, other):
        return Fraction(self.n * other.d - self.d * other.n,
                        self.d * other.d)
    def __eq__(self, other):
        self.reduce()
        other.reduce()
        return self.n == other.n and self.d == other.d
    def __gt__(self, other):
        return self.value() > other.value()
```

```
def __lt__(self, other):
    return self.value() < other.value()
```

Listing 6.7 oop_operator.py File, Fraction Class, Part 2 of 2

The b3 = b1 * b2 statement calls the special method __mul__() for object b1, passes the b2 parameter, and assigns the result to the b3 variable. This statement therefore corresponds to the statement b3 = b1.__mul__(b2). Accordingly, the __truediv__() method is called for the mathematically correct division b1 / b2; the __add__() method, for b1 + b2; and the __sub__() method, for b1 - b2.

The various calculations are carried out in the methods. The other variable refers to the transferred parameter, in this case, the Fraction object b2. A reference to a newly created Fraction object is returned to the caller. This Fraction object contains the result of the respective method, which was determined according to the rules of mathematical fraction calculation.

For b1 == b4, the __eq__() method is called. In this method, both Fraction objects are first reduced. The program then checks whether both the numerator and the denominator match.

For b1 > b4, the __gt__() method is called; for b1 < b4, the __lt__() method. In both methods, the mathematical values of the two Fraction objects are compared with each other.

6.1.6 Reference, Identity, and Copy

As described earlier in Chapter 4, Section 4.8, the name of an object is merely a reference to the object. Even with the objects of your own classes, assigning this reference to a different name simply creates a second reference to the same object.

You can create real copies of objects of your own classes by creating a new object to which you assign the values of another object of the same class.

You can also use the deepcopy() function from the copy module to copy large objects (see also Chapter 4, Section 4.8.3). Our next program contains some examples.

```
import copy

class Vehicle:
    def __init__(self, n, s):
        self.name = n
        self.speed = s
    def accelerate(self, value):
        self.speed += value
    def __str__(self):
        return f"{self.name} {self.speed} mph"
```

```
ford = Vehicle("Ford Escape", 40)
print("Original:", ford)
print()

ford2 = Vehicle(ford.name, ford.speed)
ford2.accelerate(30)
print("Copy:", ford2)
print("Identity:", ford is ford2)
print()

ford3 = copy.deepcopy(ford)
ford3.accelerate(35)
print("Copy:", ford3)
print("Identity:", ford is ford3)
print()

ford4 = ford
ford4.accelerate(40)
print("Reference:", ford4)
print("Identity:", ford is ford4)
```

Listing 6.8 oop_copy.py File

Here's the output of the program:

```
Original: Ford Escape 40 mph

Copy: Ford Escape 70 mph
Identity: False

Copy: Ford Escape 75 mph
Identity: False

Reference: Ford Escape 80 mph
Identity: True
```

The ford object is the original. The ford2 object is a copy. This new object is created using the data from the original object. The ford3 object is also a copy; it is created using the copy.deepcopy() function.

ford4 is an additional reference to the original ford object. Changing a property via this reference changes the original object, as you can see in the output.

The identity comparisons using the is operator show that only the ford and ford4 objects are identical.

6.1.7   Concatenation

A method can return the self reference (i.e., a reference to the current object). This reference is necessary if the concatenation of methods is to be enabled. This concatenation of methods is a descriptive notation for successive method calls.

As an example, consider a class for geometric rectangles that are located within a drawing area. Rectangles have a color, a position, and a size. The position is defined using the x and y coordinates of the top-left corner of the rectangle. These coordinates are measured from the top-left corner of the drawing area.

```python
class Rectangle:
    def __init__(self, c, x, y, w, h):
        self.color = c
        self.xpos = x
        self.ypos = y
        self.width = w
        self.height = h
    def __str__(self):
        return f"color: {self.color}, position: {self.xpos}/{self.ypos}" \
                f", size: {int(self.width)}/{int(self.height)}"
    def paint(self, c):
        self.color = c
        return self
    def move(self, xDelta, yDelta):
        self.xpos += xDelta
        self.ypos += yDelta
        return self
    def scale(self, xFactor, yFactor):
        self.width *= xFactor
        self.height *= yFactor
        return self

r = Rectangle("Blue", 50, 10, 30, 50)
print(r)

r.paint("Yellow")
r.move(20, 50)
r.scale(2, 2.5)
print(r)

r.paint("Blue").move(-20, -50).scale(0.5, 0.4)
print(r)
```

Listing 6.9 oop_concatenation.py File

When a rectangle object is created, all properties are assigned values. The values of the properties can be changed using the paint(), move(), and scale() methods. The paint() method receives one value with which the color is changed. The move() method receives two values with which the rectangle is moved in the x and y directions. The scale() method receives two values with which the width and height are multiplied.

Without return self, these three methods would return the value None. In this case, they could only be called one after the other.

With return self, they return a reference to the current object. This approach enables methods to be called in a chain: The painted object is moved, and the moved object is scaled.

Here's the output of the program:

```
color: Blue, position: 50/10, size: 30/50
color: Yellow, position: 70/60, size: 60/125
color: Blue, position: 50/10, size: 30/50
```

6.1.8 Nesting

The property of an object of a class can in turn be an object of another class. To clarify this correlation, now, we want to change the Vehicle class. Among other things, this class is extended by the drive property of the Engine class. Let's look at the sample program next.

```
class Engine:
    def __init__(self, p, c, f):
        self.power = p
        self.cylinder = c
        self.fuel = f

    def tune(self, x):
        self.power += x

    def __str__(self):
        return f"{self.power} / {self.cylinder} / {self.fuel}"

class Vehicle:
    def __init__(self, n, s, d):
        self.name = n
        self.speed = s
        self.drive = d

    def __str__(self):
        return f"{self.name} / {self.speed}," \
            f" Drive: {self.drive}"
```

```
ford = Vehicle("Ford SuperDuty", 70, Engine(50, 4, "Diesel"))
ford.drive.tune(10)
print(ford)

ford.drive.power = 80
ford.drive.cylinder = 6
ford.drive.fuel = "Gasoline"
print(ford)
```

Listing 6.10 oop_nesting.py File

An object of the Engine class has the power, cylinder, and fuel properties. There is also the tune() method for changing the power property and an output method.

The drive property has been added to the Vehicle class. The output method of the Vehicle class internally calls the output method of the Engine class. The output is as follows:

```
Ford SuperDuty / 70, Drive: 60 / 4 / Diesel
Ford SuperDuty / 70, Drive: 80 / 6 / Gasoline
```

An object of the Vehicle class is created in the program. The third parameter of the constructor method is an object of the Engine class.

The first dot after the name of the ford object addresses the drive property of the Vehicle object; the second dot leads to the sub-property of the Engine object.

6.1.9 Exercises

u_oop_line

Let's further develop the u_oop_point exercise from Section 6.1.4. Create a class for lines in your existing Python file whose content is supposed to be integrated. A line is marked by two points. In a graphical representation, these two points would be connected by a straight line. The class is intended to enable the following Python program, which in turn generates the next output.

```
from u_oop_geometry import *
line1 = Line()
print(line1)
point1 = Point(3.5, 2.5)
line2 = Line(Point(1.5, 4.0), point1)
print(line2)
line3 = Line(Point(-2, 5.5))
print(line3)
line4 = Line(e=Point(2.5, 1.0))
print(line4)
```

```
line4.move(-2.0, 1.5)
print(line4)
```

Listing 6.11 The u_oop_line.py File

The planned output reads as follows:

```
(0.0 / 0.0) / (0.0 / 0.0)
(1.5 / 4.0) / (3.5 / 2.5)
(-2 / 5.5) / (0.0 / 0.0)
(0.0 / 0.0) / (2.5 / 1.0)
(-2.0 / 1.5) / (0.5 / 2.5)
```

All further information on the solution can be found in the program itself and in its output.

u_oop_polygon

Now, let's develop the u_oop_point exercise from Section 6.1.4 further. Create a class for polygons in the existing Python file whose content is supposed to be integrated. A polygon is characterized by any number of points.

In a graphical representation, each point would be connected to the next point by a straight line. The last point would in turn be connected to the first point by a straight line.

This class is intended to enable the following Python program, which in turn generates the next output.

```
from u_oop_geometry import *
polygon1 = Polygon()
print(polygon1)
point1 = Point(3.5, 2.5)
point2 = Point(-2.0, 8.5)
polygon2 = Polygon([point1, Point(3.0), point2])
print(polygon2)
polygon2.move(1.0, 2.5)
print(polygon2)
```

Listing 6.12 The u_oop_polygon.py File

The planned output reads as follows:

```
(No points)
(3.5 / 2.5) / (3.0 / 0.0) / (-2.0 / 8.5)
(4.5 / 5.0) / (4.0 / 2.5) / (-1.0 / 11.0)
```

All further information on the solution can be found in the program and the output.

6.2 Advanced Topics

A class can pass its properties and methods down to another class. This mechanism is frequently used as it creates a hierarchy of classes that enables the display of objects with similar characteristics.

6.2.1 Inheritance

In the following example, a Car class is defined on the basis of the existing Vehicle class. The Car class has some additional properties and methods. With regard to the general Vehicle class, this Car class is specialized. From the perspective of the Car class, the Vehicle class is a *basic class*. From the perspective of the Vehicle class, the Car class is a *derived class*.

Let's first look at the definition of the base class Vehicle, which contains three methods and two properties.

```
class Vehicle:
    def __init__(self, n, s):
        self.name = n
        self.speed = s
    def accelerate(self, value):
        self.speed += value
    def __str__(self):
        return f"{self.name} {self.speed} mph"
```

Listing 6.13 oop_inheritance.py File, Base Class Vehicle

Now, let's look at the definition of the derived Car class.

```
class Car(Vehicle):
    def __init__(self, n, s, p):
        super().__init__(n, s)
        self.passengers = p
    def __str__(self):
        return f"{super().__str__()} {self.passengers} Passengers"
    def enter(self, number):
        self.passengers += number
    def exit(self, number):
        self.passengers -= number
```

Listing 6.14 oop_inheritance.py File, Derived Class Car

When the derived class is defined, its own name is followed by the name of the base class in parentheses.

The Car class inherits from the Vehicle class and contains five methods and three properties:

- A separate constructor method and a separate output method, each of which overwrites the method of the same name in the base class
- The custom methods enter() and exit()
- The accelerate() method inherited from the Vehicle class
- The custom passengers property
- The inherited name and speed properties

A method of a base class can be overridden in a derived class. Calling an overridden method leads to the relevant method in the derived class. If, on the other hand, you want to explicitly call the method of the next higher-level class, the super() reference is required.

Now, we've come to our main program, in which an object of the Car class is created, modified, and output.

```
fiat = Car("Fiat 500", 50, 0)
fiat.enter(3)
fiat.exit(1)
fiat.accelerate(10)
print(fiat)
```

Listing 6.15 oop_inheritance.py File, Main Program

Here's the output of the program:

```
Fiat 500 60 mph 2 Passengers
```

In general, properties and methods are first searched for in the class of the object. If they are not available in that class, the search is continued in the next higher-level class.

The fiat object is created, and the constructor is called. The constructor is found in the derived class. It passes on those properties that were not defined in the derived class by calling the constructor of the superordinate class. The super() method provides a reference to the current object in the superordinate class. Although not mandatory, this gradual approach is recommended. The remaining property of the derived class is then initialized.

The enter() and exit() methods are called. They are found in the derived class and change the passengers property.

Then, the accelerate() method is called. This method is not found in the derived class, which is why the search continues in the base class. The method is found in the base class and is used to change the speed property.

The output method is called, which is found in the derived class. It first calls the method of the same name of the superordinate class using super() so that its properties can be output. The result of this call is combined with the output of the properties of the derived class. This results in the output of all data of the object.

6.2.2   Exercise

---

u_oop_polygon_filled

Now, let's further develop the u_oop_polygon exercise from Section 6.1.9. Create a class for filled polygons in the existing Python file whose content is supposed to be integrated. A filled polygon corresponds to a polygon that is also filled with a color.

The class is intended to enable the following Python program, which in turn generates the next output.

```
from u_oop_geometry import *
polygonFilled1 = PolygonFilled([Point(3.5, 1.0),
    Point(-2.0, 6.5), Point(1.5, -3.5)], "Red")
print(polygonFilled1)
polygonFilled1.move(0.5, 2.5)
print(polygonFilled1)
polygonFilled1.paint("Blue")
print(polygonFilled1)
```

Listing 6.16 The u_oop_polygon_filled.py File

The planned output reads as follows:

```
(3.5 / 1.0) / (-2.0 / 6.5) / (1.5 / -3.5) Red
(4.0 / 3.5) / (-1.5 / 9.0) / (2.0 / -1.0) Red
(4.0 / 3.5) / (-1.5 / 9.0) / (2.0 / -1.0) Blue
```

All further information on the solution can be found in the program itself and its output.

6.2.3   Multiple Inheritance

A derived class can in turn be the base class for another class. This scenario results in inheritance across multiple levels.

With *multiple inheritance*, a class can also inherit from more than one class. In this case, it adopts the properties and methods of all base classes.

Three classes are defined in the following example:

- The MyDate class with the day, month, and year properties
- The MyTime class with the hour, minute, and second properties
- The MyDateTime class, which inherits from these two classes

An object is then created and output.

```python
class MyDate:
    def __init__(self, d, m, y):
        self.day = d
        self.month = m
        self.year = y
    def __str__(self):
        return f"{self.month}/{self.day}/{self.year}"

class MyTime:
    def __init__(self, h, m, s):
        self.hour = h
        self.minute = m
        self.second = s
    def __str__(self):
        return f"{self.hour:02d}:{self.minute:02d}:{self.second:02d}"

class MyDateTime(MyDate, MyTime):
    def __init__(self, d, mo, y, h, mi, s):
        MyDate.__init__(self, d, mo, y)
        MyTime.__init__(self, h, mi, s)
    def __str__(self):
        return f"{MyDate.__str__(self)} {MyTime.__str__(self)}"

d = MyDate(12, 8, 2024)
print(d)
t = MyTime(16, 5, 20)
print(t)
dt = MyDateTime(5, 11, 2024, 9, 35, 8)
print(dt)
```

Listing 6.17 oop_multiple.py File

Here's the output of the program:

```
8/12/2024
16:05:20
11/5/2024 09:35:08
```

The two classes from which the MyDateTime class inherits are enclosed in parentheses after the name of the class, separated by a comma.

In the MyDateTime class, the constructors or output methods of the superordinate classes can no longer be called using super() because this reference is ambiguous in this context.

However, the name of the superordinate class can also be specified explicitly. This approach makes the assignment clear. The self reference to the current object must also be noted as the first parameter.

> **Note**
>
> Not all OOP languages provide the concept of multiple inheritance, as it can make the design of class libraries complex and confusing.

6.2.4 Data Classes

Since Python 3.7, you have the option of storing related values together in a simple way using *data classes*.

Data classes are simplified classes. Some special methods are already predefined for these classes, such as a constructor method, an output method, and a method for the == operator.

In the next example, we'll define and use a data class for two-dimensional vectors.

```python
import dataclasses, math
@dataclasses.dataclass
class Vector:
    x:float = 0.0
    y:float = 0.0

    def absolute_value(self):
        return math.sqrt(self.x * self.x + self.y * self.y)

va = Vector(3.0, 4.0)
print(va)
print("Absolute value:", va.absolute_value())

vb = Vector(3.0, 4.0)
print(vb)

if va == vb:
    print("Vectors are equal")

vc = Vector(1.8)
print(vc)

vd = Vector(y=9.1)
print(vd)
```

Listing 6.18 oop_dataclass.py File

The `Vector` class is identified as a data class using the `@dataclass` decorator from the `dataclasses` module. An object of the `Vector` class has the properties `x` and `y`. These properties represent the two components of a two-dimensional vector.

The properties must be specified using type hints (see Chapter 3, Section 3.8) and can have default values. The internally predefined constructor uses the two properties as optional parameters.

The `absolute_value()` method calculates the absolute value of a two-dimensional vector based on the following mathematical rule: The amount corresponds to the square root of the total of the squares of the components.

Two objects of the `Vector` class are initialized with values for both properties. The method for the `==` operator is also predefined and compares the values of all properties.

The `vc` object receives the value for the `y` property via the default value. This principle applies to object `vd` for property `x`. The value for the `y` property is transferred here via a named parameter.

Here's the output of the program:

```
Vector(x=3.0, y=4.0)
Absolute value: 5.0
Vector(x=3.0, y=4.0)
Vectors are equal
Vector(x=1.8, y=0.0)
Vector(x=0.0, y=9.1)
```

An output method is also predefined for data classes. This output method leads to the output of the name of the class, followed by the names of the object's properties, along with their values.

6.2.5   Enumerations

Enumerations are lists of constants. The program code becomes easier to read by using the elements of an enumeration. The elements are clearly assigned to only one value. The assignment of a specific value to a second element does not make sense.

Since version 3.4, Python has provided the `IntEnum` class within the `enum` module for creating an enumeration whose elements are constants for integers.

Since Python 3.11, the `StrEnum` class is also available. The elements of this enumeration are constants for character strings. In Python 3.11 and Python 3.12, many extensions were added to enumerations, and more have already been announced for Python 3.13.

Let's start with part 1 of an example program.

```
import enum
class Color(enum.IntEnum):
    RED = 5
    YELLOW = 2
    BLUE = 4

print("Enumeration 'Color', Number:", len(Color))
for c in Color:
    print("Name:", c.name, "Value:", c.value)
c = 2
if c == Color.YELLOW:
    print("This is the YELLOW constant with the value 2")
print(Color.YELLOW)
print(Color.YELLOW * 10)
if 2 in Color:
    print("Value 2 in 'Color'")
print()
```

Listing 6.19 enumeration.py File, Part 1 of 2

The following output is generated:

```
Enumeration 'Color', Number: 3
Name: RED Value: 5
Name: YELLOW Value: 2
Name: BLUE Value: 4
This is the YELLOW constant with the value 2
2
20
Value 2 in 'Color'
```

Once the enum module has been imported, the Color enumeration is generated. This enumeration is a class that inherits from the IntEnum class. It is used to assign integer values to color names. The elements must be assigned values directly. The values do not have to be in ascending order.

The built-in len() function returns the number of elements. An enumeration can be iterated and can therefore be run through with a for loop. The name and value properties of an element provide its name and value.

The comparison of a value with other values is facilitated by these elements, which also improve the readability of your program code.

You can also incorporate the represented value into calculations. In addition, you can check whether a specific value corresponds to one of the elements.

Let's now look at part 2 of our example program.

```
class Shape(enum.IntEnum):
    CIRCLE = 3
    SQUARE = enum.auto()
    LINE = enum.auto()
    POLYGON = enum.auto()

print("Enumeration 'Shape', Number:", len(Shape))
for s in Shape:
    print("Name:", s.name, "Value:", s.value)
print()

class Weekday(enum.StrEnum):
    MONDAY = "Monday"
    TUESDAY = "Tuesday"
    WEDNESDAY = "Wednesday"

print("Enumeration 'Weekday', Number:", len(Weekday))
for w in Weekday:
    print("Name:", w.name, "Value:", w.value)
```

Listing 6.20 enumeration.py File, Part 2 of 2

This part generates the following output:

```
Enumeration 'Shape', Number: 4
Name: CIRCLE Value: 3
Name: SQUARE Value: 4
Name: LINE Value: 5
Name: POLYGON Value: 6

Enumeration 'Weekday', Number: 3
Name: MONDAY Value: Monday
Name: TUESDAY Value: Tuesday
Name: WEDNESDAY Value: Wednesday
```

The Shape enumeration is used to assign integer values to the names of geometric figures. The auto class is available in the enum module. The constructor of the auto class automatically returns integer values for the elements, always increasing by 1. If the first element were created using auto(), it would have the value 1.

The Weekday enumeration is used to assign character strings to the names of weekdays. It is a class that inherits from the StrEnum class. The properties and functions of the StrEnum class are the same as those of the IntEnum class.

6.2.6 Our Game: Object-Oriented Version

In this section, I want to present an object-oriented version of our mental calculation game. We have an object of the Game class in the application. During the lifetime of this object, several objects of the Task class are created and used.

We start with the import statement and the short main program.

```
import random
...
s = Game()
s.play()
print(s)
```

Listing 6.21 game_oop.py File, Main Program

The random module is required for the random number generator. The s object of the Game class is created in the main program. This step initializes a game. The play() method is executed for this object at the start of the process. The result is output at the end.

Let's explore the definition of the Game class.

```
class Game:
    def __init__(self):
        random.seed()
        self.correct = 0
        self.tasknumber = -1
        while self.tasknumber<1 or self.tasknumber>10:
            try:
                print("How many tasks (1 to 10):")
                self.tasknumber = int(input())
            except:
                continue

    def play(self):
        for i in range(self.tasknumber):
            a = Task(i+1, self.tasknumber)
            print(a)
            self.correct += a.answer()

    def __str__(self):
        return f"Correct: {self.correct} of {self.tasknumber}"
```

Listing 6.22 game_oop.py File, Game Class

The random generator is initialized in the constructor of the Game class. Two properties of the Game class are created (correct and tasknumber). The counter correct for correctly solved tasks is initially set to 0. The number of tasks to be solved is determined.

The desired number of tasks is generated in the play() method. Each task is an object of the Task class and is output. The answer() method is then called, i.e., the input is processed. The return value of the answer() method is 1 or 0. The counter correct is changed accordingly.

The result of the game is published in the output method.

Let's explore the definition of the Task class.

```python
class Task:
    def __init__(self, i, tasknumber):
        self.no = i
        self.tasknumber = tasknumber

    def __str__(self):
        a = random.randint(10,30)
        b = random.randint(10,30)
        self.result = a + b
        return f"Task {self.no} of {self.tasknumber}: {a} + {b}"

    def answer(self):
        try:
            if self.result == int(input()):
                print(f"{self.no}: *** Correct ***")
                return 1
            else:
                raise
        except:
            print(f"{self.no}: *** Wrong ***")
            return 0
```

Listing 6.23 game_oop.py File, Task Class

The constructor of the Task class is called with the individual number of the task and the total number of tasks. Two properties of the Task class are set (no and tasknumber).

The task is created and published in the output method. The correct result of the task is a property of the Task class.

In the answer() method, the player is asked to enter their own result, which is then compared with the correct result of the task. An incorrect or invalid entry generates an exception; this exception leads to the return of a 0. A correct entry leads to the return of a 1.

Chapter 7
Various Modules

This chapter describes programming techniques for working with time specifications, queues, collections, program parts running in parallel (multithreading), and regular expressions.

7.1   Date and Time

On many operating systems, January 1, 1970, 00:00 is the *zero point* for processing time information. Time is calculated in seconds starting from that specific point in time.

7.1.1   The time Module

The time module contains functions for saving, editing, and outputting informations about dates and times. Important functions include time(), localtime(), strftime(), and mktime().

As an alternative to these classic and frequently used functions, the datetime class from the datetime module is available (Section 7.1.4).

Our first program illustrates some options of the time module.

```python
import time
print("time():", time.time())

lt = time.localtime()
print("localtime():", f"{lt[1]:02d}/{lt[2]:02d}/{lt[0]} "
                      f"{lt[3]:02d}:{lt[4]:02d}:{lt[5]:02d}")
s = 1_720_000_000
lt = time.localtime(s)
print("localtime():", f"lt[1]:02d}/{lt[2]:02d}/{lt[0]} "
                      f"{lt[3]:02d}:{lt[4]:02d}:{lt[5]:02d}")

tu = 2024, 8, 24, 2, 35, 20, 0, 0, 0

print("strftime():", time.strftime("%m/%d/%Y %H:%M:%S"))
print("strftime():", time.strftime("%m/%d/%Y %H:%M:%S", tu))
```

```
print("strftime():", time.strftime("%m/%d/%Y %H:%M:%S", lt))

print("mktime():", time.mktime(tu))
print("mktime():", time.mktime(lt))
```

Listing 7.1 time_classical.py File

The `time()` function returns the number of seconds that have elapsed since the zero point in relation to the computer's current system time. The `localtime()` function has an optional parameter and returns a time specification as a variable of the `struct_time` type:

- If this function is called without parameters, a time is determined on the basis of the current system time of the computer.
- If this function is called with a number as a parameter, this value is added to the zero point as the number of seconds. A time is determined from this point in time.

A variable of the `struct_time` type contains a total of nine elements of a time specification, which can be accessed individually. The elements 0 to 5 contain the values for year, month, day, hour, minute and second in sequence. Element 6 contains the number of the day of the week, from Monday = 0 to Sunday = 6. The consecutive number of the day in the year is provided by element 7. Element 8 contains the daylight-saving time status: 1 = daylight-saving time is active, 0 = daylight-saving time is not active.

Here's the output of the first part of the program:

```
time(): 1709793135.3653238
localtime(): 03/07/2024 07:32:15
localtime(): 07/03/2024 11:46:40
```

The `strftime()` function is used to output a formatted time specification. This function requires a string with formatting information as the first parameter and has a second, optional parameter:

- If this function is called with only one parameter, the formatted time entry contains the current system time of the computer.
- The second parameter must be a variable of the `struct_time` type or a tuple that has the same structure as such a variable. The formatted time is derived from the elements. The day of the week, day of the year, and daylight-saving time are often unknown at first. If a 0 is simply set for each of these parameters, the correct values will be determined automatically.
- The following formatting specifications are available: %d = day, %m = month, %Y = year, %H = hour, %M = minute, and %S = second. All outputs are in 2 digits and, where applicable, with a leading zero. The 4-digit output for the year is an exception.

The mktime() function returns the number of seconds since the zero point for a time specification. This function requires a parameter, which must be a tuple of nine elements or a variable of the struct_time type. The number of seconds can be used to calculate with time specifications.

Here's the output of the second part of the program:

```
strftime(): 03/07/2024 07:32:15
strftime(): 08/24/2024 02:35:20
strftime(): 07/03/2024 11:46:40
mktime(): 1724463320.0
mktime(): 1720000000.0
```

7.1.2 Stopping a Program

The sleep() function from the time module allows you to stop a program for a certain period of time.

In our next sample program, the current time is output several times within a loop. After each output, the program is stopped for 2 seconds using the sleep() function. At the end, the time difference between the start and end is calculated.

```
import time

starttime = time.time()
print("Start:", starttime)
for i in range(5):
    time.sleep(2)
    print(time.time())
endtime = time.time()
print("End:", endtime)

difference = endtime - starttime
print("Difference:", difference)
```

Listing 7.2 time_stop.py File

The following output is generated:

```
Start: 1709793205.1911144
1709793207.2485235
1709793209.2722845
1709793211.2935443
1709793213.3188007
1709793215.3400621
End: 1709793215.3556974
Difference: 10.164582967758179
```

As the output of the difference shows, the sleep() function can only be used for time control with limited accuracy for the following reasons:

- First, the individual program steps require their own, albeit short, runtime, depending on the speed and load of the computer.
- Secondly, the time intervals differ slightly.

Nevertheless, you can use the sleep() function for many applications, including graphical animations, simulations, or game programming.

7.1.3 The timedelta Class

The timedelta class from the datetime module enables the processing of a time duration (i.e., the difference between two points in time). Calculations and comparisons can be carried out using operators. Let's consider a sample program.

```
import datetime

d1 = datetime.timedelta(5)
print("Only days:", d1)

d2 = datetime.timedelta(5, 8, 0, 0, 10, 2, 1)
print("All parameters:", d2)

d3 = datetime.timedelta(days=1, hours=30, minutes=-20, seconds=80)
print("Named parameters, conversion:", d3)
print(f"Properties: {d3.days} days, {d3.seconds} seconds")

sc = d3.total_seconds()
mi = sc / 60
st = mi / 60
print(f"Method, conversion: {sc:.0f} sec. = {mi:.3f} min. = {st:.3f} hr.")
print()

print("Operators:")
a = datetime.timedelta(days=5, hours=10)
print("Time difference a:", a)
b = datetime.timedelta(days=2, hours=4)
print("Time difference b:", b)

print("a + b:", a + b)
print("a - b:", a - b)
print("b * 2.5:", b * 2.5)
print("a / 2:", a / 2)
print("a / b:", a / b)
```

```
print("a % b:", a % b)
print("a == b:", a == b)
print("a > b:", a > b)
```

Listing 7.3 time_timedelta.py File

The constructor of the timedelta class has a total of seven optional parameters; their names in order are days, seconds, microseconds, milliseconds, minutes, hours, and weeks.

The first timedelta object is only created with the value 5. Therefore, it has the value of 5 days. The second timedelta object is created with all values in the correct order. Named parameters are used with the creation of the third timedelta object.

If the numerical range of a time specification is exceeded (or fallen short of) during the creation of an object, the value will be converted accordingly:

- One day and 30 hours becomes 2 days and 6 hours.
- 6 hours and –20 minutes become 5 hours and 40 minutes.
- 40 minutes and 80 seconds become 41 minutes and 20 seconds.

Next follows two properties and a method of the timedelta object:

- days contains only the days.
- seconds contains only the seconds in the day, without the days themselves.
- total_seconds() returns the total number of seconds.

In the program, the latter value is converted into minutes and hours and output with formatting.

Two time differences can be added together and subtracted from each other. A time difference can be multiplied by a numerical value or divided by a numerical value. The division of two time differences results in a numerical value. The remainder of an integer division of two time differences results in a time difference. Two time differences can be compared with each other.

The following output is generated:

```
Only days: 5 days, 0:00:00
All parameters: 12 days, 2:10:08
Named parameters, conversion: 2 days, 5:41:20
Properties: 2 days, 20480 seconds
Method, conversion: 193280 sec. = 3221.333 min. = 53.689 hr.


Operators:
Time difference a: 5 days, 10:00:00
Time difference b: 2 days, 4:00:00
a + b: 7 days, 14:00:00
a - b: 3 days, 6:00:00
```

```
b * 2.5: 5 days, 10:00:00
a / 2: 2 days, 17:00:00
a / b: 2.5
a % b: 1 day, 2:00:00
a == b: False
a > b: True
```

7.1.4 The datetime Class

The datetime class from the datetime module is used to store, edit, and display information about dates and times. Let's look at an example program.

```python
import datetime

dt1 = datetime.datetime(2024, 7, 23, 9, 28, 7, 500000)
print("Time:", dt1)

dt2 = datetime.datetime.fromisoformat("2024-07-23T09:28:07")
print("From ISO format:", dt2)
print("Formatted:", dt2.strftime("%m/%d/%Y %H:%M:%S"))
print(f"Day: {dt2.day}, Month: {dt2.month}, Year: {dt2.year}, " \
      f"Hour: {dt2.hour}, Minute: {dt2.minute}, Second: {dt2.second}")
print()

dt2 = dt2.replace(month=9, day=14, minute=35, second=40)
print("Replace:", dt2)
print()

print("dt1 == dt2:", dt1 == dt2)
print("dt1 < dt2:", dt1 < dt2)
print()

td1 = datetime.timedelta(days=3, hours=17)
dt3 = dt2 + td1
print("Add time difference:", dt3)
td2 = dt3 - dt2
print("Time difference:", td2)
print()

dt4 = datetime.datetime.now()
print("Now:", dt4)
wdays = ["", "Monday", "Tuesday", "Wednesday",
         "Thursday", "Friday", "Saturday", "Sunday"]
```

```
wd = dt4.isoweekday()
print(f"Weekday: {wd} = {wdays[wd]}")
```

Listing 7.4 time_datetime.py File

The constructor of the datetime class has a total of ten parameters. The first three are mandatory; the rest are optional. The names of the first seven parameters in order are year, month, day, hour, minute, second, and microsecond. The most important properties of a datetime object have the same names.

The static fromisoformat() method enables you to create a datetime object from a string that follows the International Organization for Standardization (ISO) 8601 format. The number of possible formats has been extended with Python 3.11.

The strftime() method is used to individually format the time specification (see also Section 7.1.1).

You can use the replace() method to replace individual parts of a time specification with new values. The names of the named parameters in turn correspond to the properties of a datetime object.

Two datetime objects can be compared using operators.

A timedelta object can be added to or subtracted from a datetime object. The result is another datetime object. If you subtract a datetime object from another datetime object, you get a timedelta object.

The static now() method creates a datetime object with the current time. The isoweekday() method returns the number of the weekday of a datetime object. The days of the week are counted from Monday = 1 to Sunday = 7.

The output of the program reads as follows:

```
Time: 2024-07-23 09:28:07.500000
From ISO format: 2024-07-23 09:28:07
Formatted: 07/23/2024 09:28:07
Day: 23, Month: 7, Year: 2024, Hour: 9, Minute: 28, Second: 7

Replace: 2024-09-14 09:35:40

dt1 == dt2: False
dt1 < dt2: True

Add time difference: 2024-09-18 02:35:40
Time difference: 3 days, 17:00:00

Now: 2024-03-07 07:34:56.130043
Weekday: 4 = Thursday
```

225

7.1.5   The calendar Class

The calendar class from the calendar module provides useful functions in connection with calendar dates. Our next sample program illustrates a few of its numerous methods.

```python
import calendar, datetime

print("Leap year:", calendar.isleap(2024))
print("Number of leap years:", calendar.leapdays(2020, 2030))
print()

calendar.prmonth(2024, 2)
print()

print("Full weeks:")
cal = calendar.Calendar()
it = cal.itermonthdates(2024, 2)
for date in it:
    print(date.strftime("%m/%d"), end=" ")
    if date.isoweekday() == 7:
        print()
print()

print("Enumeration via weekdays:")
for day in calendar.Day:
    print(day, end=" ")
print()

if calendar.weekday(2024,2,1) == calendar.THURSDAY:
    print("2/1/2024 is a Thursday")
print()

print("Enumeration via months:")
for month in calendar.Month:
    print(month, end=" ")
print()

d = datetime.datetime(2024,2,1)
if d.month == calendar.FEBRUARY:
    print("Month is February")
```

Listing 7.5 time_calendar.py File

The static `isleap()` method determines whether the year in question is a leap year. The static `leapdays()` method returns the number of leap years for a period that is specified using two year values.

The static `prmonth()` method outputs a calendar for a specific month in a table. The table has a heading. Above the individual columns are the names of the days of the week in abbreviated form. Each row contains the days of a week, starting with Monday. All weeks containing days of the month in question are displayed.

The `itermonthdates()` method can be called for an object of the `calendar` class. The year and month of a specific month are passed as parameters. The method returns an iterator over all days of the full weeks that contain days of the relevant month.

The iterator can be used to control a `for` loop. Each element that is passed through using the iterator corresponds to an object of the `date` class from the `datetime` module. The `date` class corresponds to the `datetime` class from Section 7.1.4, limited to the date information (that is to say, without the time information).

The individual elements are output formatted in the `for` loop. The `isoweekday()` method is used to determine the number of the weekday for a `datetime` object, with Monday = 1 to Sunday = 7. A line break is added after every Sunday.

The `calendar.day` enumeration has been available since Python 3.12. This enumeration contains the names and values of the days of the week, from MONDAY = 0 to SUNDAY = 6. The individual elements can also be accessed directly, from `calendar.MONDAY` through `calendar.SUNDAY`.

You can use the static `weekday()` method of the `calendar` class to determine the number of the day of the week for a specific date, with Monday = 0 to Sunday = 6. The program uses the `calendar.THURSDAY` constant to determine whether the date in question is a Thursday.

The `calendar.Month` enumeration has also been available since Python 3.12. It contains the names and values of the months, from JANUARY = 1 to DECEMBER = 12. The individual elements can also be accessed directly, from `calendar.JANUARY` through `calendar.DECEMBER`.

The `month` property of the `datetime` class provides the number of the month for a specific date, from JANUARY = 1 to DECEMBER = 12. The program uses the `calendar.FEBRUARY` constant to determine whether the date in question is in February.

Here's the output of the program:

```
Leap year: True
Number of leap years: 3

   February 2024
Mo Tu We Th Fr Sa Su
         1  2  3  4
```

```
 5  6  7  8  9 10 11
12 13 14 15 16 17 18
19 20 21 22 23 24 25
26 27 28 29

Full weeks:
01/29 01/30 01/31 02/01 02/02 02/03 02/04
02/05 02/06 02/07 02/08 02/09 02/10 02/11
02/12 02/13 02/14 02/15 02/16 02/17 02/18
02/19 02/20 02/21 02/22 02/23 02/24 02/25
02/26 02/27 02/28 02/29 03/01 03/02 03/03

Enumeration via weekdays:
0 1 2 3 4 5 6
2/1/2024 is a Thursday

Enumeration via months:
1 2 3 4 5 6 7 8 9 10 11 12
Month is February
```

7.1.6   Our Game: Version with Time Measurement

Time specifications enable us to extend our mental calculation game. Now, we can measure the time it takes to solve one or all tasks. If the best performance is achieved, for example, a high number of tasks solved in the first attempt or solving the tasks within a particularly short time, we can record this achievement with the date and time.

In this version of our mental calculation game, we have five addition problems with numbers from 10 to 30. Only one attempt is allowed per task. Invalid entries or incorrect results are simply commented on with the text Wrong. The total time required is shown at the end.

```
import random, time

random.seed()
correct = 0
starttime = time.time()

for tasknumber in range(5):
    a = random.randint(10,30)
    b = random.randint(10,30)
    c = a + b
```

```
    print(f"Task {tasknumber+1} of 5: {a} + {b}")
    try:
        number = int(input("Please enter suggested solution: "))
        if number == c:
            print("Correct")
            correct += 1
        else:
            raise
    except:
        print("Wrong")

difference = time.time() - starttime
print(f"Correct: {correct} of 5 in {difference:.2f} seconds")
print("Result achieved: ", time.strftime("%m/%d/%Y %H:%M:%S"))
```

Listing 7.6 game_time.py File

Some invalid or incorrect results are also entered. A possible output could read as follows:

```
Task 1 of 5: 23 + 28
Please enter a suggested solution: 51
Correct
Task 2 of 5: 22 + 28
Please enter a suggested solution: 50
Correct
Task 3 of 5: 19 + 14
Please enter a suggested solution: 33
Correct
Task 4 of 5: 16 + 19
Please enter a suggested solution: 32
Wrong
Task 5 of 5: 23 + 30
Please enter a suggested solution: 32
Wrong
Correct: 3 of 5 in 11.56 seconds
Result achieved: 03/07/2024 09:35:28
```

The time is taken at the beginning and again at the end, and the difference is then calculated. The correct results are counted. In the event of an invalid input or an incorrect result, an exception is generated, and the text Wrong is displayed.

229

7.1.7   Our Game: Object-Oriented Version with Time Measurement

The object-oriented version of our mental calculation game with time measurement is based on the object-oriented version without time measurement (see the *game_oop.py* file in Chapter 6, Section 6.2.6). In this section, I only present the extensions. First, let's look at the import statements and the short main program.

```python
import random, time

# Main program
s = Game()
s.measure(True)
s.play()
s.measure(False)
print(s)
```

Listing 7.7 game_time_oop.py File, Main Program

In the main program, the time module is required for time measurement. The measure() method is called before and after the game to determine the duration of the game. The bool type parameter specifies whether it is the start or the end of the game.

The definition of the Task class is not changed. Our new definition of the Game class adds the measure() method and changes the output method.

```python
class Game:
...
    def measure(self, start):
        if start:
            self.starttime = time.time()
        else:
            self.time = time.time() - self.starttime

    def __str__(self):
        d = time.strftime("%m/%d/%Y")
        t = time.strftime("%H:%M:%S")
        return f"Correct: {self.correct} of {self.tasknumber} " \
               f"in {self.time:.2f} seconds\non {d} at {t}"
```

Listing 7.8 game_time_oop.py File, Game Class, Excerpt

The start time is determined in the measure() method if the parameter value True is transmitted. Otherwise, the end time and the duration of the game will be determined. The playing time is a property of the Game class. The result of the game is published in the output method together with the duration of the game as well as the date and time.

7.2   Queues

In general, a *queue* always arises if a system receives more requests within a period of time than it can process. This can be a queue of people at a movie theater box office or a queue of sent data in a network.

The aim is to reduce the queue as quickly as possible. Depending on the problem, this goal can be achieved in few different ways:

- In a *first in, first out (FIFO) queue*, the request that arrives first is also processed first. Requests can only arrive at the end of the queue and can only be processed at the beginning of the queue.

- In a *last in, first out (LIFO) queue*, the request that arrived last is processed first. Requests can only arrive at the end of the queue and can only be processed there.

- A *double-ended queue* has no beginning, but two ends. Requests can arrive at both ends of the queue and can also be processed at both ends.

The queue and collections modules provide some specialized data types for processing queues. They are similar to the built-in data types (such as dictionary, list, set, or tuple), but these special data types have additional capabilities and allow fast access.

7.2.1   The SimpleQueue Class

Since Python 3.7, the queue module contains the SimpleQueue class for processing a FIFO queue. Let's consider an example.

```
import queue
x = queue.SimpleQueue()
x.put(5)
x.put(19)
x.put(-2)
x.put(12)
print("Number of elements:", x.qsize())

while not x.empty():
    print(x.get())
print("Number of elements:", x.qsize())
```

Listing 7.9 queue_fifo.py File

The x object of the SimpleQueue class is created. The qsize() method returns the number of elements. The put() method adds individual elements to the end of the queue. The empty() method returns information as to whether the queue is empty. The get() method removes an element from the beginning of the queue and returns its value at the same time.

Here's the output of the program:

```
Number of elements: 4
5
19
-2
12
Number of elements: 0
```

The value 5 was the first to be added to the queue. It is also the first value to be removed (following FIFO).

7.2.2 The LifoQueue Class

To process a LIFO queue, the queue module provides the LifoQueue class. Let's look at an example.

```
import queue
x = queue.LifoQueue()
x.put(5)
x.put(19)
x.put(-2)
x.put(12)
print("Number of elements:", x.qsize())

while not x.empty():
    print(x.get())
print("Number of elements:", x.qsize())
```

Listing 7.10 queue_lifo.py File

The get() method removes an element from the end of the queue. The elements are therefore output in reverse order:

```
Number of elements: 4
12
-2
19
5
Number of elements: 0
```

7.2.3 The PriorityQueue Class

The PriorityQueue class from the queue module provides a special feature. The elements can be added to the queue unsorted. The get() method removes the elements from the queue based on an ascending sort order, either by number or by character. If the elements are of different types, an exception will occur.

Let's consider an example.

```python
import queue
x = queue.PriorityQueue()
x.put(5)
x.put(19)
x.put(-2)
x.put(12)
print("Number of elements:", x.qsize())

while not x.empty():
    print(x.get())
print("Number of elements:", x.qsize())
```

Listing 7.11 queue_sorted.py File

The output reads as follows:

```
Number of elements: 4
-2
5
12
19
Number of elements: 0
```

7.2.4   The deque Class

The deque (pronounced "deck") class is available in the collections module for processing a double-ended queue. This class provides far more options than the previously discussed queue classes. In our first example, some operations are performed on a double-ended queue.

```python
import collections

d1 = collections.deque([8, 18, 28])
print("Created:", d1)
d2 = d1.copy()
print("Copy:", d2)

print("Elements:")
for x in d1:
    print(x)
```

```
for i in range(len(d1)):
    print(f"{i}: {d1[i]}")

d3 = d1 * 2
print("Deque doubled:", d3)
d4 = collections.deque([9, 19, 29])
d5 = d1 + d4
print("Other deque added:", d5)

d1.clear()
print("After emptying:", d1)
```

Listing 7.12 deque_operation.py File

The following output is generated:

```
Created: deque([8, 18, 28])
Copy: deque([8, 18, 28])
Elements:
8
18
28
0: 8
1: 18
2: 28
Deque doubled: deque([8, 18, 28, 8, 18, 28])
Other deque added: deque([8, 18, 28, 9, 19, 29])
After emptying: deque([])
```

The deque() function returns a new object of the deque data type. An iterable is required as a parameter. Since Python 3.5, you can create a copy using the copy() method. You can access the individual elements using a for loop with or without an index.

Since Python 3.5, you can also multiply a double-ended queue using the * operator and add other double-ended queues using the + operator. The clear() method deletes all elements.

In this next example, a double-ended queue is changed in various ways.

```
import collections
d = collections.deque([8, 18, 28])
print("Created:", d)

d.appendleft(5)
print("Element appended to the left:", d)
d.append(25)
```

```
print("Element appended to the right:", d)
d.insert(2, 11)
print("Element inserted:", d)
d.extendleft([7,9])
print("Elements appended to the left:", d)
d.extend([17,19])
print("Elements appended to the right:", d)

x = 5
if x in d:
    print(f"Position of {x}: {d.index(5)}")

for i in range(5):
    le = d.popleft()
    print("Element removed on the left:", le)
ri = d.pop()
print("Element removed on the right:", ri)
print("After:", d)

d.rotate()
print("After rotation by +1:", d)
d.rotate(-1)
print("After rotation by -1:", d)
```

Listing 7.13 deque_change.py File

The following output is generated:

```
Created: deque([8, 18, 28])
Element appended to the left: deque([5, 8, 18, 28])
Element appended to the right: deque([5, 8, 18, 28, 25])
Element inserted: deque([5, 8, 11, 18, 28, 25])
Elements appended to the left: deque([9, 7, 5, 8, 11, 18, 28, 25])
Elements appended to the right: deque([9, 7, 5, 8, 11, 18, 28, 25, 17, 19])
Position of 5: 2
Element removed on the left: 9
Element removed on the left: 7
Element removed on the left: 5
Element removed on the left: 8
Element removed on the left: 11
Element removed on the right: 19
After: deque([18, 28, 25, 17])
After rotation by +1: deque([17, 18, 28, 25])
After rotation by -1: deque([18, 28, 25, 17])
```

You can use the `appendleft()` and `append()` methods to append individual elements to the left or right end.

Since Python 3.5, with the `insert()` method, you can insert an element at the specified position. If the position does not exist, the element will be appended to the right end.

The `extendleft()` and `extend()` methods are used to extend multiple elements that originate from an iterable at the left or right end. The elements are added one after the other in the order of the iterable. For this reason, the order for the added elements in the `extendleft()` method will change.

Since Python 3.5, you can use the `index()` method to search for the position of a specific value. If there is no element with this value, an exception will be generated.

The `popleft()` and `pop()` methods remove an element at the left or right end and return the removed element as the return value.

You can use the `rotate()` method to rotate the elements circularly. If you do not specify a parameter, the system rotates one element to the right. If you enter negative values, the system rotates to the left. An element that is pushed over the edge during rotation will be reattached at the other end.

7.3 Multithreading

The term *multithreading* refers to the ability of a program to process multiple parts—the *threads*—in parallel. The Python programming language provides the routines for multithreading.

7.3.1 What Is Multithreading Used For?

A number of problems exist in which multithreading proves useful, for example, in simulations of real processes in which one action triggers another action, but then continues to run itself. The triggered action can in turn trigger a third action and so on. All actions access the same data and change it if necessary. They end at different times and with different results.

Threads can also be used within graphical user interface (GUI) applications in which computationally intensive program parts run in the background (*are swapped out*) or the results of measured value evaluations or other real-time processes are displayed in parallel.

7.3.2 Generating a Thread

The `threading` module provides a simple option for multithreading.

In the following program, a subthread is started from the main program, the main thread. The main thread must then wait ten seconds before it terminates. The start and end of the main thread are displayed.

In the subthread, the thread_process() function runs during the waiting time of the main thread, in which the current time is displayed five times in total. After each display, the subthread is paused for 1.5 seconds. The start and end of the subthread are also displayed.

```
import time, threading

def thread_process():
    print("Start subthread")
    for i in range(5):
        print(i, time.time())
        time.sleep(1.5)
    print("End of subthread")

print("Start main thread")
t = threading.Thread(target=thread_process)
t.start()
time.sleep(10)
print("End of main thread")
```

Listing 7.14 thread_generate.py File

The following output is generated:

```
Start main thread
Start subthread
0 1709800916.7200632
1 1709800918.244327
2 1709800919.7767298
3 1709800921.3086305
4 1709800922.842534
End of subthread
End of main thread
```

To generate a subthread, an object of the Thread class is created in the main thread. The constructor expects at least the named target parameter, which transmits the name of a callback function that is run through when the subthread is started. The subthread is started by calling the thread_process() function for the thread object. With the times given, the main thread does not end until the subthread has ended.

7.3.3   Identifying a Thread

To differentiate between the effects of different threads, each thread has a unique identification number (*ID*), which you can determine using the get_ident() function. Because the main program is also a thread, it has an ID, too.

Now, we want to expand on the previous program. From the main thread, two subthreads are started at different times, each using the thread_process() function. The ID is determined for each thread and also displayed for each output.

```python
import time, threading

def thread_process():
    id = threading.get_ident()
    print("Start subthread", id)
    for i in range(5):
        print(i, "Subthread", id)
        time.sleep(1.5)
    print("End of subthread", id)
    return

id = threading.get_ident()
print("Start main thread", id)

t1 = threading.Thread(target=thread_process)
t1.start()
time.sleep(0.5)
t2 = threading.Thread(target=thread_process)
t2.start()

time.sleep(10)
print("End of main thread", id)
```

Listing 7.15 thread_ident.py File

The program generates the following output, for example:

```
Start main thread 10844
Start subthread 7764
0 Subthread 7764
Start subthread 8436
0 Subthread 8436
1 Subthread 7764
1 Subthread 8436
2 Subthread 7764
2 Subthread 8436
```

```
3 Subthread 7764
3 Subthread 8436
4 Subthread 7764
4 Subthread 8436
End of subthread 7764
End of subthread 8436
End of main thread 10844
```

The main thread in this example has the ID 10844, the first subthread has the ID 7764, and the second subthread has the ID 8436. You can observe the parallel running of the threads based on their IDs.

7.3.4   Shared Data and Objects

All subthreads have access to all global data and objects of the main thread. In this example, a specific variable is shared and changed by the main thread and two subthreads.

```python
import time, threading

def thread_process():
    global counter
    id = threading.get_ident()
    for i in range(5):
        counter += 1
        print(i, id, counter)
        time.sleep(1.5)
    return

id = threading.get_ident()
counter = 0
print(id, counter)

t1 = threading.Thread(target=thread_process)
t1.start()
time.sleep(0.5)
t2 = threading.Thread(target=thread_process)
t2.start()
time.sleep(10)

counter += 1
print(id, counter)
```

Listing 7.16 thread_common.py File

The following example output is generated:

```
8844 0
0 6792 1
0 10444 2
1 6792 3
1 10444 4
2 6792 5
2 10444 6
3 6792 7
3 10444 8
4 6792 9
4 10444 10
8844 11
```

The counter variable is global in the main thread (in our case, ID 8844) and is preset with 0. It is made known in the thread function using the global keyword (see also Chapter 5, Section 5.6.6), incremented by 1, and then displayed. In both subthreads (in our case, ID 6792 and ID 10444), the variable is known and can be changed. At the end, it is also increased by 1 and displayed in the main thread.

7.3.5 Threads and Exceptions

If an unhandled exception occurs, it only affects the thread in which it occurs. Neither the calling main thread nor the other subthreads are disturbed by this exception.

Let's now expand the previous program again. If the counter variable has the value 5, the unhandled ZeroDivisionError exception will be generated artificially. The subthread in which this happens will be closed immediately. The other threads—including the main thread—will continue unaffected. Let's look now at our amended program.

```
import time, threading

def thread_process():
    global counter
    id = threading.get_ident()
    for i in range(5):
        counter += 1
        print(i, id, counter)
        if counter == 5:
            res = 1/0
        time.sleep(1.5)
    return


id = threading.get_ident()
```

```
counter = 0
print(id, counter)

t1 = threading.Thread(target=thread_process)
t1.start()
time.sleep(0.5)
t2 = threading.Thread(target=thread_process)
t2.start()
time.sleep(10)

counter += 1
print(id, counter)
```

Listing 7.17 thread_exception.py File

The following example output is generated:

```
5704 0
0 5196 1
0 11268 2
1 5196 3
1 11268 4
2 5196 5
Exception in thread Thread-1 (thread_process):
Traceback (most recent call last):
...
ZeroDivisionError: division by zero
2 11268 6
3 11268 7
4 11268 8
5704 9
```

The subthread with ID 5196 terminates due to the unhandled exception. The subthread with ID 11268 ends regularly, as does the main thread with ID 5704.

7.4   Regular Expressions

I now want to provide insight into the extensive possibilities of *regular expressions*. They are used to search and change character strings via *search patterns* and are used in many programming languages; see *https://en.wikipedia.org/wiki/Regular_expression*.

Regular expressions are often used to check and evaluate entries or to search for files. If the program determines that an incorrect or incomplete entry has been made, the system can provide assistance and prompt you for a new entry.

In Python, the re module allows you to use regular expressions. The two programs we'll write in this section demonstrate some of its extensive possibilities.

7.4.1 Finding Text Parts

In the next program, the findall() function is used to search for parts of texts. It finds all text segments that match the regular expression and returns a list of these text parts for further evaluation. The program is subdivided for a better overview. In addition, the individual blocks of the program and their associated outputs are numbered.

Let's look at the first part of this program.

```
import re

tx = "house and mouse and louse"
print(tx)
print("1:", re.findall("mouse",tx))
print("2:", re.findall("[hm]ouse",tx))
print("3:", re.findall("[l-m]ouse",tx))
print("4:", re.findall("[^l-m]ouse",tx))
print("5:", re.findall(".ouse",tx))
print("6:", re.findall("^.ouse",tx))
print("7:", re.findall(".ouse$",tx))
print()
...
```

Listing 7.18 regexp_search.py File, Part 1

The following output, including the numbers, results from the first part of this program:

```
house and mouse and louse
1: ['mouse']
2: ['house', 'mouse']
3: ['mouse', 'louse']
4: ['house']
5: ['house', 'mouse', 'louse']
6: ['house']
7: ['louse']
```

The first seven regular expressions refer to the house and mouse and louse text:

❶ The mouse string is searched for. All occurrences of this character string are returned.

❷ The texts parts that begin with an h or an m and end with ouse are searched for. In the rectangular brackets, we are basically saying "Search for h or m".

❸ The text parts that begin with one of the characters from l to m and end with ouse are searched for. A range is specified in the rectangular brackets using the hyphen.

❹ The text parts that *do not* begin with one of the characters from l to m but end with ouse are searched for. The ^ character represents a logical negation in the context of a range.

❺ The system searches for text parts that begin with any character and end with ouse. The . character means "any character".

❻ As before, but only valid for text parts that appear at the *beginning* of the text being analyzed. The ^ character means "at the beginning".

❼ As before, but only valid for text parts that are *at the end* of the text being analyzed. The $ character means "at the end".

Let's now consider the second part of this program.

```
...
tx = "0172-445633"
print(tx)
print("8:", re.findall("[0-2]",tx))
print("9:", re.findall("[^0-2]",tx))
print("10:", re.findall("[047-]",tx))
print()
...
```

Listing 7.19 regexp_search.py File, Part 2

The following output results from the second part of this program:

```
0172-445633
8: ['0', '1', '2']
9: ['7', '-', '4', '4', '5', '6', '3', '3']
10: ['0', '7', '-', '4', '4']
```

The next three expressions refer to the text 0172-445633:

❽ One of the digits from 0 to 2 is searched for as a text segment. This time, the range includes digits.

❾ A character that is *not* in the number range from 0 to 2 is searched for as a text part. All digits from 3 and all non-digits are found.

❿ One of the characters from the specified set of characters is searched for as a text part. All characters or digits mentioned are found.

Finally, here is the last part of this program:

```
...
tx = "aa and aba and abba and abbba and aca"
print(tx)
```

```
print("11:", re.findall("ab*a",tx))
print("12:", re.findall("ab+a",tx))
print("13:", re.findall("ab?a",tx))
print("14:", re.findall("ab{2,3}a",tx))
```

Listing 7.20 regexp_search.py File, Part 3

The output of this final part of the program reads as follows:

```
aa and aba and abba and abbba and aca
11: ['aa', 'aba', 'abba', 'abbba']
12: ['aba', 'abba', 'abbba']
13: ['aa', 'aba']
14: ['abba', 'abbba']
```

The last four regular expressions refer to the text, aa and aba and abba and abbba and aca. This text illustrates the behavior regarding the number of occurrences of certain characters.

⓫ All text parts are found that contain, in sequence, an a, any number (0 is also possible) of the character b, again an a. The * character (asterisk) means "any number of occurrences of the specific character".

⓬ All text parts are found that contain an a, at least one b, again an a. The + (plus) sign means "The number of occurrences of the specific character is 1 or greater".

⓭ All text parts are found that contain an a, none or one b, again an a. The question mark ? means in this context "The number of occurrences of the specified character is 0 or 1".

⓮ All text parts are found that contain an a, two or three times the character b, again an a. The curly brackets allow you to specify the desired number of occurrences.

7.4.2 Replacing Text Parts

The sub() function replaces all text parts that match the regular expression with another text. For ease of understanding, let's use the same texts and regular expressions in this next program as in the previous program.

```
import re

tx = "house and mouse and louse"
print(tx)
print("1:", re.sub("mouse","x",tx))
print("2:", re.sub("[h|m]ouse","x",tx))
print("3:", re.sub("[l-m]ouse","x",tx))
print("4:", re.sub("[^l-m]ouse","x",tx))
print("5:", re.sub(".ouse","x",tx))
```

```
print("6:", re.sub("^.ouse","x",tx))
print("7:", re.sub(".ouse$","x",tx))
print()

tx = "0172-445633"
print(tx)
print("8:", re.sub("[0-2]","x",tx))
print("9:", re.sub("[^0-2]","x",tx))
print("10:", re.sub("[047-]","x",tx))
print()

tx = "aa and aba and abba and abbba and aca"
print(tx)
print("11:", re.sub("ab*a","x",tx))
print("12:", re.sub("ab+a","x",tx))
print("13:", re.sub("ab?a","x",tx))
print("14:", re.sub("ab{2,3}a","x",tx))
```

Listing 7.21 regexp_replace.py File

The output of this program reads as follows:

```
house and mouse and louse
1: house and x and louse
2: x and x and louse
3: house and x and x
4: x and mouse and louse
5: x and x and x
6: x and mouse and louse
7: house and mouse and x


0172-445633
8: xx7x-445633
9: 01x2xxxxxxx
10: x1x2xxx5633

aa and aba and abba and abbba and aca
11: x and x and x and x and aca
12: aa and x and x and x and aca
13: x and x and abba and abbba and aca
14: aa and aba and x and x and aca
```

All text parts found are replaced by x for clarification.

7.5   Audio Output

The winsound module allows you to output system sounds and WAV files via your speakers on Windows. Let's consider a small sample program.

```
import winsound, time
for i in range(600, 1500, 200):
    winsound.Beep(i, 500)

time.sleep(3)
winsound.PlaySound("SystemQuestion", winsound.SND_ALIAS)
winsound.PlaySound("GChord.wav", winsound.SND_FILENAME)
print("End")
```

Listing 7.22 sound.py File

The Beep() function sends a sound signal. The first parameter represents the frequency, the second parameter the duration of the sound signal.

The PlaySound() function can be used to play Windows system sounds or WAV files. The first parameter is the name of the system sound or the name of the WAV file. The second parameter is a constant. In the case of a system sound, SND_ALIAS must be specified. For a WAV file, you would use SND_FILENAME.

Chapter 8, Section 8.9, shows another example of using the winsound module. There, a text is converted into Morse code, which is output as an audio signal via a loudspeaker.

Differences on Ubuntu Linux and macOS

The winsound module is not available.

Chapter 8
Files

Data can be stored permanently in simple files or in databases. In this chapter, you'll learn about various techniques for saving data in files.

Toward the end of the chapter, we'll work on a version of our mental calculation game that allows the player's name and the time required to be saved permanently in a file.

8.1 Opening and Closing a File

A file that is supposed to be edited must first be opened, which is done using the built-in open() function. For this function to work, you'll need to specify the name of the file (possibly with its path) and the opening mode. The return value is a file object that is used for further access to the file.

In the programs in this section, we assumed that the file to be opened is located in the same directory as the Python program. If it isn't, you must specify the absolute or relative path to the file in accordance with Table 8.1. A file can be opened in different modes, listed in Table 8.2.

Description	Path
To the *sample.txt* file in the *sub* subdirectory	sub/sample.txt
To the *sample.txt* file in the parent directory	../sample.txt
To the *sample.txt* file in the parallel *par* directory	../par/sample.txt
To the *sample.txt* file in the *C:\Temp* directory	C:/Temp/sample.txt

Table 8.1 Relative and Absolute Path Information

Note

In Python on Windows, both the forward slash / and the backslash \ are permitted for directory changes. In this book, I mostly use the UNIX-friendly variant of the forward slash /.

Mode	Description
r	For reading. As it is the default value, this parameter can be omitted.
w	For writing.
a	For appending to the end of the file.
r+	For reading and writing, with the current access position moved to the beginning.
w+	For writing. The file is emptied first.
a+	For reading and writing with the current access position moved to the end.
b	For opening a file in binary mode that is to be read or written. This option is an additional specification, see also Section 8.4.

Table 8.2 Opening Modes

The current access position is saved in each file object. This position indicates where reading or writing is currently taking place. The current access position changes with every read or write operation, but you can also be change it using the seek() method without having to read and write.

After editing, a file must be closed using the close() method. If it is not closed, the file may be blocked from further access.

8.2   Text Files

Text files can be written or read sequentially. Their lines vary in length and usually end with a line break. The content of the text file can be edited using a simple editor. Direct access to a specific line using a program is not possible as the length of the lines is not known.

8.2.1   Writing a Text File

To write to a file, it must first be opened using the open() function and the opening mode w. If the file does not yet exist, it will be created at that very moment. Careful: If it already exists, it will be overwritten without warning!

A runtime error may occur when you open a file in write mode (for example, when writing to a write-protected medium or to a non-existent directory). You should therefore perform exception handling, which terminates the program using the exit() function from the sys module if necessary.

If the file opens successfully, you can write individual strings to it using the write() method and an iterable with character strings using the writelines() method. You must add the \n character (*new line*) for each line end.

```
import sys

try:
    f = open("write.txt","w")
except:
    print("File not opened")
    sys.exit(0)

f.write("The first line\n")
for i in range(2,11,2):
    f.write(f"{i} ")
f.write("\n")

li = ["Houston", "Boston", "Miami"]
f.writelines(li)
f.write("\n")
for i in li:
    f.write(f"{i}\n")

dc = {"Peter":31, "Judith":28, "William":35}
f.writelines(dc)
f.write("\n")
for k,v in dc.items():
    f.write(f"{k}:{v}\n")

f.close()
```

Listing 8.1 write.py File

The *write.txt* file now contains the following content:

```
The first line
2 4 6 8 10
HoustonBostonMiami
Houston
Boston
Miami
PeterJudithWilliam
Peter:31
Judith:28
William:35
```

Listing 8.2 write.txt File

First, a single line of text is written to the file using the write() method. The numbers in the sequence 2, 4, 6, 8, 10 are then written to a line in the file using a formatted string literal. An end-of-line character is also output after both lines.

A list of character strings is then written to the file twice:

1. As a whole using the writelines() method where the elements appear in one line without separation between the elements

2. Element by element using a for loop and the write() method and with an end-of-line character in each case

Finally, a dictionary in which the keys are character strings is written twice to the file:

1. As a whole using the writelines() method where only the keys appear in one line, again without separation

2. Element by element using a for loop through the items view of the dictionary, the write() method, and with an end-of-line character in each case

> **Note**
>
> If you use the opening mode "a" instead of "w" when opening the file (f = open ("write.txt", "a")), the output will be appended to the previous file content. The file would then get bigger and bigger.

> **Differences on Ubuntu Linux and macOS**
>
> Creating files using a Python program may fail if you do not have the necessary permissions.

8.2.2   Reading a Text File

You can read the lines of a text file using the following methods, among others:

- readlines() reads all lines into a list of strings.
- readline() reads one line into a string.
- read() reads all lines into a string.

Let's look at the content of the *read.txt* file for a better comparison of these three methods. Among other things, the file contains some figures. The total of these figures is calculated and output.

```
6
2.5
abc
-4
```

Listing 8.3 read.txt File

Let's look at this program next.

```python
import sys

def open_file():
    global f
    try:
        f = open("read.txt")
    except:
        print("File not opened")
        sys.exit(0)

def value(line):
    try:
        return float(line)
    except:
        return 0.0

open_file()
li = f.readlines()
total = 0
for line in li:
    total += value(line)
print("Total:", total)
f.close()

open_file()
total = 0
line = f.readline()
while line:
    total += value(line)
    line = f.readline()
print("Total:", total)
f.close()

open_file()
tx = f.read()
li = tx.split("\n")
total = 0
for line in li:
    total += value(line)
print("Total:", total)
f.close()
```

Listing 8.4 read.py File

The file is opened for reading in the open() function. The global f variable is a reference to the file object that is required to access the content. If the file cannot be opened, for example, because it does not exist in the indicated directory, the program terminates.

In the value() function, the numerical value of a character string is determined and returned. If the conversion is not successful, the value 0.0 will be returned, which does not affect the totaling.

If the three reading methods are used, the file is opened at the beginning and closed again at the end.

The list of character strings returned by the readlines() method is run through using a for loop. The numerical value of each character string is determined and added to the total.

The readline() method is first called individually, then in a while loop. Each time the method is called, it returns a string whose numerical value is determined and added to the total. When the end of the file is reached, the string is empty and thus ends the while loop.

> **Note**
>
> A (visually) empty line still contains the character for the end of the line, so the line itself is not empty from the perspective of the readline() method.

The string returned by the read() method is split into a list of strings using the split() method and the end-of-line character. This list is run through using a for loop. The numerical value of each character string is determined and added to the total.

The output of the program reads:

```
Total: 4.5
Total: 4.5
Total: 4.5
```

8.2.3   Writing a CSV File

Comma-separated values (CSV) files are text files that contain one data record per line. The data within a data record is separated by a defined separator character, often a semicolon or a comma. CSV files can be read by many programs (such as Microsoft Excel or LibreOffice Calc).

In addition to the examples in this section, you'll find another version of our mental calculation game in Section 8.10. The name of the player and the time it took them are saved in a CSV file, which can be used as the basis for a highscore list.

We now want to save a table with personal data consisting of three data records in a CSV file. Each data record contains a person's family name, first name, personnel number, salary, and date of birth. The table is saved in the program in a two-dimensional list.

```python
import sys
try:
    f = open_file("data.csv","w")
except:
    print("File not opened")
    sys.exit(0)

li = [["Mayer", "John", 6714, 3500, "3/15/1962"],
      ["Smith", "Peter", 81343, 3750, "4/12/1958"],
      ["Mertins", "Judith", 2297, 3621.5, "12/30/1959"]]
for ds in li:
    f.write(f"{ds[0]};{ds[1]};{ds[2]};{ds[3]};{ds[4]}\n")
f.close()
```

Listing 8.5 write_csv.py File

The output file *data.csv* then contains the following content:

```
Mayer;John;6714;3500;3/15/1962
Smith;Peter;81343;3750;4/12/1958
Mertins;Judith;2297;3621.5;12/30/1959
```

Listing 8.6 data.csv File

After opening the file, the multidimensional list is created. It is run through using a for loop. With each run, the data of a data record is written to the file using the write() method. A semicolon is inserted between each element.

To ensure successful export, the structure of the data must be known. The data of the different data types must be processed individually.

If Microsoft Excel is installed on Windows with the default settings, double-clicking on the *data.csv* output file is sufficient to display the content as a table in Microsoft Excel, as shown in Figure 8.1.

	A	B	C	D	E
1	Mayer	John	6714	3500	3/15/1962
2	Smith	Peter	81343	3750	4/12/1958
3	Mertins	Judith	2297	3621.5	12/30/1959

Figure 8.1 CSV File in Microsoft Excel

Differences on Ubuntu Linux and macOS

On Ubuntu Linux and macOS you can open the file using LibreOffice Calc. Double-click on the filename in the directory display to open the **Text import** dialog box.

Make sure that in the **Separator options** area, only the options **Separated by** and **Semicolon** are checked. If you also used the **Comma** option, the numbers with decimal places would be split into two columns. The data is then displayed correctly as in Microsoft Excel on Windows.

If necessary, in the **Character set** in the list, select the **System** entry.

8.2.4 Reading a CSV File

To successfully import a CSV file, the structure and data types of the data must be known. Prior to running the following example, a data record was added to the CSV file from the previous example in Microsoft Excel, as shown in Figure 8.2.

	A	B	C	D	E
1	Mayer	John	6714	3500	3/15/1962
2	Smith	Peter	81343	3750	4/12/1958
3	Mertins	Judith	2297	3621.5	12/30/1959
4	Weaver	Jeremy	4711	2900	8/12/1976

Figure 8.2 CSV File in Microsoft Excel, Supplemented

The file is read using our next example program. The data in the data records is output individually.

```python
import sys
try:
    f = open_file("data.csv")
except:
    print("File not opened")
    sys.exit(0)

tx = f.read()
f.close()
li = tx.split("\n")

for line in li:
    if line:
        ds = line.split(";")
        print(f"{ds[0]} {ds[1]} {ds[2]} {ds[3]} {ds[4]}")
```

Listing 8.7 read_csv.py File

The program outputs all four data records:

```
Mayer John 6714 3500 3/15/1962
Smith Peter 81343 3750 4/12/1958
Mertins Judith 2297 3621.5 12/30/1959
Weaver Jeremy 4711 2900 8/12/1976
```

First, the entire text of the file is saved in a string using the read() method. This string is split into a list of strings using the split() method based on the end-of-line character. Each string contains a line from the file (i.e., a data record). Further processing will only be carried out if the line is not empty.

Each line is split, again using the split() method, into a list with the individual data of the data record using the semicolon.

Finally, the data is output, separated from each other by blank spaces.

8.3   Files with a Defined Structure

If a file is *formatted*, its structure is fixed. The file contains data records with a defined length and structure. Line breaks may or may not exist.

As a result, you know exactly what information is in which place. The required information can be read directly from the file without having to read the entire file line by line.

You can also change certain information *selectively* without affecting the remaining content of the file. Both options are presented in the following sections.

8.3.1   Formatted Writing

Formatted writing to a file resembles a formatted output on the screen. Data records of the same length and structure are created. Optionally, you can insert an end-of-line character after each data record to make the file easier to read in an editor. You can also change the file using an editor as long as you adhere to the defined structure.

In the following program, the table of people from Section 8.2.3 is formatted and saved in a file.

```python
import sys
try:
    f = open_file("formatted.txt","w")
except:
    print("File not opened")
    sys.exit(0)

li = [["Mayer", "John", 6714, 3500, "3/15/1962"],
      ["Smith","Peter", 81343, 3750, "4/12/1958"],
```

```
    ["Mertins", "Judith", 2297, 3621.5, "12/30/1959"]]
for ds in li:
    f.write(f"{ds[0]:12}{ds[1]:12}{ds[2]:6}{ds[3]:8.2f}{ds[4]:>11}\n")
f.close()
```

Listing 8.8 write_formatted.py File

The *formatted.txt* file is shown in Figure 8.3.

Figure 8.3 formatted.txt File

The file is now opened for writing. The data records are output using formatted string literals, as described in Chapter 5, Section 5.2.2. In addition, \n is added at the end of each line.

Each data record has the same size of 51 bytes on Windows. It results from the total of the individual formats and the 2 bytes for the end of the line: 12 + 12 + 6 + 8 + 11 + 2 = 51. The entire file has a size of 51 × 3 = 153 bytes.

> **Differences on Ubuntu Linux and macOS**
>
> The character for the end of a line has a size of 1 byte. For this reason, each data record has 50 bytes, and the file has 150 bytes.

8.3.2   Reading in Any Position

The seek() method allows you to change the current read or write position within the file. It can be called with one or two parameters. If this method is called with one parameter, this parameter indicates the distance in bytes, measured from the start of the file. The start of the file corresponds to the value 0.

A desired position is reached directly without reading the information before the position in question. You can then read or write from the position reached.

The seek() method is introduced in our next example, together with a variant of the read() method. This method also allows a certain number of bytes to be read.

```
import sys
try:
    f = open_file("formatted.txt")
except:
```

```
    print("File not opened")
    sys.exit(0)

dslength = 51
for i in range(3):
    f.seek(dslength*i)
    name = f.read(12).strip()
    f.seek(30 + dslength*i)
    salary = float(f.read(8))
    print(name, salary)

f.close()
```

Listing 8.9 read_any.py File

The output reads:

```
Mayer 3500.0
Smith 3750.0
Mertins 3621.5
```

First, the file is opened for reading. Positions in all data records are then reached one after the other using the seek() method:

- At positions 0 × 51 = 0, 1 × 51 = 51 and 2 × 51 = 102, the next 12 bytes are read with the family name of the respective person. The strip() method is used to remove the spaces after the name.

- At positions (0 × 51 + 30) = 30, (1 × 51 + 30) = 81 and (2 × 51 + 30) = 132, the next 8 bytes are read with the salary of the respective person. The float() function is used to convert what is read into a number.

> **Differences on Ubuntu Linux and macOS**
>
> Because the end-of-line character is 1 byte, dslength must have the value 50.

8.3.3   Writing in Any Position

You can open a file in r+ mode, and then you can both read and change it. In the following program, the salaries of the persons are read, changed, and then saved again.

```
import sys
try:
    f = open_file("formatted.txt", "r+")
except:
    print("File not opened")
```

```
    sys.exit(0)

dslength = 51
for i in range(3):
    f.seek(30 + dslength*i)
    salary = float(f.read(8))
    salary = round(salary * 1.05, 2)
    # salary = round(salary / 1.05, 2)
    f.seek(30 + dslength*i)
    f.write(f"{salary:8.2f}")

f.close()
```

Listing 8.10 write_any.py File

The *formatted.txt* file is shown in Figure 8.4.

```
formatted.txt - Notepad
File   Edit   Format   View   Help
Mayer          John          6714 3675.00  3/15/1962
Smith          Peter        81343 3937.50  4/12/1958
Mertins        Judith        2297 3802.58 12/30/1959
```

Figure 8.4 formatted.txt File after Modification

The salary of each person is read as in the previous program. It is then increased by 5%. You could also use the commented-out line to reduce salaries by 5% instead. The value is rounded to two decimal places. The position at which the salary is located is reached again by using seek(). The changed salary is then formatted and written to the file.

Differences on Ubuntu Linux and macOS

Because the end-of-line character is 1 byte, dslength must again have the value 50.

8.4   Serialization Using pickle

The pickle module is used to serialize and deserialize objects using Python. These objects can be objects of the built-in data types or objects of your own classes.

During serialization, objects are saved as a byte sequence in a file. During deserialization, the objects are reloaded from a file. When saving, the type of object is added so that it is known when the object is loaded.

In addition to the examples in this section, you'll find another version of our mental calculation game in Section 8.10. The player's name and time required are saved permanently using serialization so that a highscore list can be maintained.

8.4.1 Writing Objects to Files

You can use the dump() function to serialize objects. In our next example, several objects are saved, one after the other, in a file.

```python
import pickle, sys
class Vehicle:
    def __init__(self, n, s):
        self.name = n
        self.speed = s
    def __str__(self):
        return f"{self.name} {self.speed} mph"

try:
    f = open_file("pickle_file.bin", "wb")
except:
    print("File not opened")
    sys.exit(0)

tu = [4,"abc",8], -12.5, {"a":1, "b":1}, set([8, 5, 2, 5])
pickle.dump(tu, f)

ford = Vehicle("Ford Escape", 40)
pickle.dump(ford, f)

pickle.dump(3, f)
pickle.dump("Boston", f)
pickle.dump("Houston", f)
pickle.dump("Detroit", f)

f.close()
```

Listing 8.11 pickle_write.py File

The program starts with the familiar definition of the Vehicle class. The file is opened in wb mode (i.e., for writing in binary mode). I have also selected the *.bin* file extension.

A tuple is created that consists of a list, a number, a dictionary, and a set. The list contains, among other things, a character string. The tuple is saved in binary format in the file using the dump() function. The first parameter is the name of the saved object; the second parameter is the file object.

An object of the Vehicle class is then created and also saved in binary format in the file.

If you want to save a specific number of objects and restore them later, for example, three character strings, you should first save this number in binary format in the file and then the objects themselves. In this way, the correct number of objects can be reloaded later (see also Section 8.4.2).

Note

If the file already exists, it will be overwritten. To append objects, you can open the file in ab mode. The file can only be written or read using the methods described here.

8.4.2   Reading Objects from Files

Objects that have been saved in a file using the dump() function can be retrieved from the file in the same order by using the load() function. If an object is of a separate class, its definition must be known.

In our next program, objects that were serialized using our previous program are deserialized again and are thus transferred.

```python
import pickle, sys
class Vehicle:
    def __init__(self, n, s):
        self.name = n
        self.speed = s
    def __str__(self):
        return f"{self.name} {self.speed} mph"

try:
    f = open_file("pickle_file.bin", "rb")
except:
    print("File not opened")
    sys.exit(0)

tu = pickle.load(f)
print(tu)

ford = pickle.load(f)
ford.speed += 20
print(ford)
```

```
number = pickle.load(f)
for i in range(number):
    print(pickle.load(f))

f.close()
```

Listing 8.12 pickle_read.py File

The following output is generated:

```
([4, 'abc', 8], -12.5, {'a': 1, 'b': 1}, {8, 2, 5})
Ford Escape 60 mph
Boston
Houston
Detroit
```

The program starts with the definition of the custom Vehicle class. The file is opened in rb mode (i.e., for reading in binary mode).

The tuple is first loaded using the load() function from the pickle module. All elements of the tuple are recognized correctly together with their data types. The next step is to load the object of the Vehicle class. You can see from the change that this object and its class have also been recognized correctly.

The number of subsequent objects is then loaded as a number. In other words, the number of objects to be loaded for the for loop is known. All objects are displayed on the screen for checking.

8.5 Data Exchange Using JSON

JavaScript Object Notation (JSON) is a widely used format for exchanging data. Data in the JSON format can be read in a text editor. You can use different programming languages to edit JSON content.

A single object of one of the following data types can be saved directly in JSON format: number, string, list, tuple, and dictionary. The following special features must be observed:

- A tuple is saved as a list but can then be generated again from this list.
- A set can't be saved directly in JSON format but can be saved using a list, for example.
- An object of a custom data type can be saved in JSON format using the special __dict__ property and then recreated.

8.5.1 Writing Objects to Files

In the following program, a tuple is created that consists of a list, a number, a dictionary, a set, and an object of a custom class. The list contains, among other things, a character string. The tuple is then saved in JSON format in a file.

```python
import json, sys

class Vehicle:
    def __init__(self, n, s):
        self.name = n
        self.speed = s
    def __str__(self):
        return f"{self.name} {self.speed} mph"

try:
    f = open_file("json_file.json", "w")
except:
    print("File not opened")
    sys.exit(0)

s = set([8, 5, 2, 5])
ls = []
for element in s:
    ls.append(element)

ford = Vehicle("Ford Escape", 40)

tu = [4,"abc",8], -12.5, {"a":1, "b":1}, ls, ford.__dict__
json.dump(tu, f)
f.close()
```

Listing 8.13 json_write.py File

The elements of the set are saved in a list prior to being added to the tuple. Instead of the object of a custom class, its special property __dict__ is stored in the tuple. The dump() function from the json module writes the tuple as a single object in JSON format to a file that was previously opened for writing. The last step is to close the file, which now contains the following content:

```
[[4, "abc", 8], -12.5, {"a": 1, "b": 1}, [8, 2, 5],
    {'name': "Ford Escape", "speed": 40}]
```

Listing 8.14 json_file.json File

8.5.2 Reading Objects from Files

In our next program, a tuple is read from a file as a single JSON object and then output.

```python
import json, sys

class Vehicle:
    def __init__(self, n, s):
        self.name = n
        self.speed = s
    def __str__(self):
        return f"{self.name} {self.speed} mph"

try:
    f = open_file("file.json", "r")
except:
    print("File not opened")
    sys.exit(0)

li = json.load(f)
f.close()

s = set()
for element in li[3]:
    s.add(element)

ford = Vehicle(li[4]["name"], li[4]["speed"])

tu = li[0], li[1], li[2], s, ford.__str__()
print(tu)
```

Listing 8.15 json_read.py File

The load() function from the json module reads a JSON object from a file that was previously opened for reading. Next, the file is closed. The JSON object contains a list whose elements are used to create the original tuple from the previous program.

The set is initially empty. The elements of the sub-list created from the original set are added to it. The object of the custom class is created using the constructor from the dictionary that was created using the special __dict__ property. This is followed by the output of the tuple:

```
([4, 'abc', 8], -12.5, {'a': 1, 'b': 1}, {8, 2, 5}, 'Ford Escape 40 mph')
```

8.6 Editing Multiple Files

Previously, only a single file whose name was known was ever processed. If you're faced with the task of processing multiple files from a specific directory whose number and names are unknown, you can use the glob() function from the glob module and the scandir() function from the os module. They each provide a list that enables access to the individual files.

8.6.1 The glob.glob() Function

In the following example, the names of specific files from the current directory are first saved in a list using the glob() function. These files are opened and searched one after the other. If a specific search text is found in one of the files, its name will be displayed.

```python
import glob
file_list = glob.glob("wri*.py")
for file in file_list:
    try:
        f = open(file)
    except:
        print("File not opened")
        continue
    totaltext = f.read()
    f.close()
    if totaltext.find("Smith") != -1:
        print(file)
```

Listing 8.16 file_list.py File

The following output is generated:

```
write_csv.py
write_formatted.py
```

The parameter of the glob() function is a character string that serves as a filter for the file search. The string can also contain one of the wildcards, that is, * (for any number of arbitrary characters) or ? (for a single arbitrary character). This search string searches for files whose name starts with wri and ends with .py.

Each element of the generated list corresponds to a file name. Each of these files is opened. If opening was not successful, the system continues with the next file. If the opening was successful, the complete content of that file is read into a character string using the read() method. The file is then closed again.

The string is searched using the find() method. If the search text is found, the file name is displayed.

Since Python 3.5, you can also search an entire directory tree recursively using the glob.glob() function and the ** string. The following examples provide lists of all files with the extension .*py* whose name starts with c, but in different locations:

- glob.glob("c*.py") only runs through the current directory.
- glob.glob("**/c*.py") only runs through direct subdirectories.
- glob.glob("**/c*.py", recursive=True) runs through the current directory, direct subdirectories, their subdirectories etc. (i.e., the entire directory tree).

8.6.2 The os.scandir() Function

8

Since Python 3.5, the scandir() function from the os module has provided an alternative to traversing a directory. Our next program provides the same result as our previous program.

```python
import os
entry_iterator = os.scandir(".")
for entry in entry_iterator:
    if entry.name.startswith("wri") and entry.name.endswith(".py"):
        try:
            f = open(entry)
        except:
            print("File not opened")
            continue

        totaltext = f.read()
        f.close()
        if totaltext.find("Smith") != -1:
            print(entry.name)
entry_iterator.close()
```

Listing 8.17 file_list_scandir.py File

The scandir() function expects the name of a path as a parameter. You can use a period (.) to specify the current directory. The function returns an iterator of type nt.Scandir-Iterator, which is used to iterate through all entries in the directory.

A single entry in the directory has the nt.DirEntry type. Its name property contains its name. If the name begins with wri and ends with .py, the respective file will be processed as described earlier.

Since Python 3.6, you have the option of closing the iterator using close(). This step releases the resources used at the same time.

8.7 Information about Files

The stat() function from the os module provides a range of information about a single file. Let's look at an example.

```python
import os, time
info = os.stat("formatted.txt")
print("Size in bytes:", info.st_size)
fm = "%m/%d/%Y %H:%M:%S"
print("Last access:", time.strftime(fm, time.localtime(info.st_atime)))
print("Last modification:", time.strftime(fm, time.localtime(info.st_mtime)))
print("First creation:", time.strftime(fm, time.localtime(info.st_birthtime)))
```

Listing 8.18 file_info.py File

The following output is generated:

```
Size in bytes: 153
Last access: 03/07/2024 09:51:25
Last modification: 03/02/2024 17:29:00
First creation: 02/28/2024 16:15:11
```

The return value of the stat() function is an object of the stat_result type. It has the following properties, among others:

- st_size: Size in bytes
- st_atime: Time of last access
- st_mtime: Time of last change
- st_birthtime: Time of first creation (new in Python 3.12)

The time information is converted using the localtime() function, and the output is formatted using the strftime() function. As the format is used multiple times, it should be saved in a variable.

Differences on Ubuntu Linux and macOS
The data records only have a length of 50 characters instead of 51. This results in a file size of 150 bytes.

8.8 Managing Files and Directories

The os and shutil modules provide additional functions for managing files and directories. In our next example, files are copied, renamed, and deleted.

```
import sys, shutil, os, glob
print(glob.glob("re*.txt"))

if not os.path.exists("read.txt"):
    print("File does not exist")
    sys.exit(0)

shutil.copy("read.txt","read_copy.txt")
print(glob.glob("re*.txt"))

try:
    shutil.move("read_copy.txt","read_new.txt")
except:
    print("Error when renaming")
print(glob.glob("re*.txt"))

try:
    os.remove("read_new.txt")
except:
    print("Error when removing")
print(glob.glob("re*.txt"))
```

Listing 8.19 file_manage.py File

On Windows, the following output is generated:

```
['read.txt']
['read.txt', 'read_copy.txt']
['read.txt', 'read_new.txt']
['read.txt']
```

The first line of the output provides the original list of files that start with re and end with .txt.

The exists() function from the os.path module is used to check whether a specific file exists that is supposed to be copied. If the file does not exist, the program will terminate prematurely.

The *read.txt* file is then copied using the copy() function from the shutil module. The copy is called *read_copy.txt*. The second line of the output provides the modified list of files.

The *read_copy.txt* file is then renamed to *read_new.txt* using the move() function from the shutil module. The next line of the output again provides the changed list of files. The move() function can also be used to move a file to a different directory.

Finally, the *read_copy.txt* file is deleted using the remove() function from the os module. The final line of the output provides the final list of files.

8.9   The Morse Code Sample Project

This section presents a sample project in which files, external modules, strings, dictionaries, some functions, and the winsound module for sound output (see Chapter 7, Section 7.5) are used.

A sample text is converted into Morse code. The code consists of individual Morse code characters and pauses. A Morse code character represents a single character of a text and is made up of a sequence of short and long signals. Texts have been transmitted as sequences of sound, radio, or light signals for a long time through Morse code.

8.9.1   Reading Morse Code from a File

First, the various characters and the corresponding Morse code are read from the *morse.txt* file into a dictionary. In the file, a short signal is represented by a dot; a long signal, by a hyphen. The individual characters and the corresponding Morse code character are each separated by a space.

```
A .-
B -...
C -.-.
D -..
E .
```

Listing 8.20 morse.txt File, Excerpt

The function readCode() in our morse module (*morse.py*) is used for reading. The module can be made available to other Python programs.

```
import sys
def readCode():
    try:
        f = open("morse.txt")
    except:
        print("File not opened")
        sys.exit(0)
    all_lines = f.readlines()
    f.close()
    code = {}
    for line in all_lines:
        words = line.split()
        code[words[0]] = words[1]
```

```
    for i in range(97,123):
        code[chr(i)] = code[chr(i-32)]
    return code
```

Listing 8.21 morse.py File

In the readCode() function, all lines of the *morse.txt* file are read first. Then, an empty dictionary is created.

Each line is split into two-character strings using the space character. The first character string contains a Unicode character, and the second character string contains the corresponding Morse code character. The Unicode character is used as the key for the Morse code character entry in the dictionary.

Morse code is not case sensitive. For this reason, the dictionary elements for the lower-case letters are assigned the same Morse code as the corresponding elements for the uppercase letters.

The dictionary is returned as the result of the function.

8.9.2   Output on the Screen

In this step, the Morse code characters of a sample text are displayed on the screen. The individual Morse code characters are separated by a space. Let's look at this program now.

```
import sys, morse
def writeCode(text, code):
    for character in text:
        try:
            print(code[character], end=" ")
        except KeyError:
            print(" ", end=" ")
    print()

code = morse.readCode()
writeCode("Hello World", code)
```

Listing 8.22 morse_screen.py File

The output of the "Hello World" sample text in Morse code looks as follows:

.... . .¯.. .¯.. ---   .¯¯ --- .¯. .¯.. ¯..

In the main program, the readCode() function from the morse module described earlier is called. It returns a dictionary. The writeCode() function is then called with a sample text and this dictionary.

In the writeCode() function, a passed sample text is encoded in Morse code and output. If there is no equivalent to a character in the dictionary, a KeyError will occur, and a space will be output.

8.9.3   Output with Sound Signals

Now the Morse code characters of a sample text are output as audio signals. Before you start the following program, you should switch on your Windows PC speakers:

```python
import sys, morse, time, winsound
def soundCode(text, code):
    signalDuration = {".":200, "-":600}
    signalPause = 0.2
    characterPause = 0.6
    wordPause = 1.4

    allWords = text.split()
    for w in range(len(allWords)):
        word = allWords[w]
        for c in range(len(word)):
            character = word[c]
            print(character, end="")
            try:
                allSignals = code[character]
                for s in range(len(allSignals)):
                    signal = allSignals[s]
                    winsound.Beep(800, signalDuration[signal])
                    if s < len(allSignals)-1:
                        time.sleep(signalPause)
                if c < len(word)-1:
                    time.sleep(characterPause)
            except KeyError:
                pass
        if w < len(allWords)-1:
            print(" ", end="")
            time.sleep(wordPause)

code = morse.readCode()
soundCode("Hello World", code)
```

Listing 8.23 morse_sound.py File

In the soundCode() function, a passed sample text is encoded in Morse code and output as a sequence of sound signals. Another dictionary is created first, which sets the signal duration for a short signal at 200 milliseconds. Since a long signal must be three times as long, it therefore has a length of 600 milliseconds.

Each signal is followed by a signal pause lasting one short signal, each character is followed by a character pause lasting one long signal, and each word is followed by a word pause lasting seven short signals.

The entire text is broken down into individual words using the spaces. Each word is broken down into individual characters. The corresponding entry in the Morse code dictionary is searched for each character. Each Morse code character found is broken down into individual signals. Each signal is output as a sound using the Beep() function from the winsound module.

If no equivalent to a character exists in the dictionary, a KeyError will occur. In this case, the character is ignored.

The sleep() function from the time module takes care of the signal, character, and word pauses.

The readCode() function from the morse module is called in the main program. This function returns a dictionary. The soundCode() function is then called with a sample text and this dictionary.

> **Differences on Ubuntu Linux and macOS**
>
> The winsound module is not available.

8.10   Our Game: Version with Highscore File

What follows now is another version of our mental calculation game in the *game_file.py* file. The name and time required are saved permanently using serialization (Section 8.4) so that a highscore list can be maintained.

The name of the player is requested first. Then, five addition problems with numbers from 10 to 30 are set. Only one attempt is available per task. Invalid entries or incorrect results are commented using the text Wrong. Finally, the total time required for the solution attempts is given.

If all five tasks are solved correctly, the name and time taken are recorded in a highscore list, which is saved in a file.

8.10.1   Sample Entries

As shown in Figure 8.5, you can see a typical run of the program.

First, the highscores are considered. There are already some entries there. Then a round is played in which all five tasks are solved correctly. This changes the highscore. The last step is to exit the program.

```
Please enter (0: End, 1: Highscores, 2: Play): 1
 P.  Name              Time
 1.  Susan          10.07 sec
 2.  Ralph          10.11 sec
 3.  Paul           12.42 sec
Please enter (0: End, 1: Highscores, 2: Play): 2
Please enter your name (max. 10 characters): Judith
Task 1 of 5: 13 + 21 : 34
*** Correct ***
Task 2 of 5: 10 + 20 : 30
*** Correct ***
Task 3 of 5: 18 + 29 : 47
*** Correct ***
Task 4 of 5: 26 + 19 : 45
*** Correct ***
Task 5 of 5: 27 + 12 : 39
*** Correct ***
Result: 5 of 5 in 8.86 seconds, Highscore
 P.  Name                Time
 1.  Judith          8.86 sec
 2.  Susan          10.07 sec
 3.  Ralph          10.11 sec
 4.  Paul           12.42 sec
Please enter (0: End, 1: Highscores, 2: Play):
```

Figure 8.5 Sample Entries

8.10.2   Structure of the Program

The program consists of a main program and four functions. In the main program, a main menu is called within an endless loop. The highscores can be displayed, the game can be played, or the program can be closed.

The four functions of the program are:

- hs_read(): Reads highscores from the file into a list
- hs_show(): Displays the highscore list on the screen
- hs_write(): Writes the highscore list to the file
- game(): Sets the tasks, solves the tasks, adds the recorded time to the highscore list

8.10.3   The Code of the Program

The following listing shows the start of the program, together with the function for reading the highscore data from the file:

```
import random, time, glob, pickle

def hs_read():
    global hs_list
```

```
    if not glob.glob("highscore.bin"):
        hs_list = []
        return
    f = open("highscore.bin", "rb")
    hs_list = pickle.load(f)
    f.close()
```

Listing 8.24 game_file.py File, Part 1 of 5

The modules random (to generate the random numbers), time (to measure the time), glob (to check the file), and pickle (to access the file) are required.

After reading from the file, the names and the required times appear in the hs_list list. As the list is used for the first time in this function, but is required throughout the entire program, it is made known here via global in the global namespace.

If the file does not exist, an empty list is generated. If the file does exist, the entire content is read into the hs_list list using the load() function from the pickle module.

This is followed by the function for displaying the highscore:

```
def hs_show():
    if not hs_list:
        print("No highscores available")
        return
    print(" P. Name                Time")
    for i in range(len(hs_list)):
        print(f"{i+1:2d}. {hs_list[i][0]:10} {hs_list[i][1]:5.2f} sec")
        if i >= 9:
            break
```

Listing 8.25 game_file.py File, Part 2 of 5

If the highscore list is empty, a corresponding message will be displayed. If the list is not empty, a maximum of ten highscores is displayed formatted with the information on *rank*, *name*, and *time* after a heading.

This is followed by the function for writing the highscore list to the file:

```
def hs_write():
    f = open("highscore.bin","wb")
    pickle.dump(hs_list,f)
    f.close()
```

Listing 8.26 game_file.py File, Part 3 of 5

The entire list is written to the file using the dump() function from the pickle module.

This is followed by the game function:

```python
def game():
    name = input("Please enter your name (max. 10 characters): ")
    name = name[:10]
    correct = 0
    starttime = time.time()

    for tasknumber in range(5):
        a = random.randint(10,30)
        b = random.randint(10,30)
        c = a + b

        try:
            number = int(input(f"Task {tasknumber+1} of 5: {a} + {b} : "))
            if number == c:
                print("*** Correct ***")
                correct += 1
            else:
                raise
        except:
            print("* Wrong *")

    difference = time.time() - starttime
    print(f"Result: {correct:d} of 5 in {difference:.2f} seconds", end = "")
    if correct == 5:
        print(", Highscore")
    else:
        print(", Sorry, no highscore")
        return

    found = False
    for i in range(len(hs_list)):
        if difference < hs_list[i][1]:
            hs_list.insert(i, [name, difference])
            found = True
            break
    if not found:
        hs_list.append([name, difference])
    hs_show()
```

Listing 8.27 game_file.py File, Part 4 of 5

Once the name has been entered, it is shortened to a maximum of ten characters for uniform output. Then follow some program parts that are already known from earlier versions of the game, such as setting the five tasks, entering the solutions, evaluating the solutions, measuring the time.

If not all five tasks are solved correctly, no entry can be made in the highscore list.

The highscore list is searched element by element. If an element is found in which a time greater than the new result time is entered, the new result time is inserted into the highscore list at this point using the insert() method.

If no entry was made by inserting, the new result time is appended to the end of the list using the append() method. This ensures that the highscore list is always sorted in ascending order according to score time. The list is displayed after each change.

Here's the main program with the menu:

```
# Program
random.seed()
hs_read()

while True:
    try:
        menu = int(input("Please enter "
            "(0: End, 1: Highscores, 2: Play): "))
    except:
        print("Wrong input")
        continue

    if menu == 0:
        break
    elif menu == 1:
        hs_show()
    elif menu == 2:
        game()
    else:
        print("Wrong input")

hs_write()
```

Listing 8.28 game_file.py File, Part 5 of 5

When the random generator has been initialized, the highscore file is loaded. The endless loop starts with the main menu. You can enter "0", "1" or "2"; all other entries are wrong.

- When you enter "1", the highscore list is displayed and then the main menu appears again.
- When you enter "2", the game runs, and the main menu appears again.
- When you enter "0", the endless loop is exited. The highscore list is written to the file. The program ends.

8.11   Our Game: Object-Oriented Version with Highscore File

The following is another object-oriented version of the mental calculation game in the *game_file_oop.py* file. This time, the name and time required are saved in a CSV file, which you can view using Microsoft Excel on Windows or LibreOffice Calc on Ubuntu Linux and macOS.

In addition to the Game and Task classes, we also need the Highscore class. The Task class has not changed compared to the version in the *game_oop.py* file.

First, we'll tackle the import statement and the main program.

```
import random, time, glob
...
# Program
while True:
    try:
        menu = int(input("Please enter "
            "(0: End, 1: Highscores, 2: Play): "))
    except:
        print("Wrong input")
        continue

    if menu == 0:
        break
    elif menu == 1:
        hs = Highscore()
        print(hs)
    elif menu == 2:
        s = Game()
        s.measure(True)
        s.play()
        s.measure(False)
        print(s)
    else:
        print("Wrong input")
```

Listing 8.29 game_file_oop.py File, Main Program

The modules random (to generate the random numbers), time (to measure the time), and glob (to check the file) are required.

To view the highscore, an object of the Highscore class is created and output. When a game is being played, an object of the Game class is created. After measuring the time and playing the game, the results are displayed.

Now, let's look at the Game class.

```python
class Game:
    def __init__(self):
        random.seed()
        self.correct = 0
        self.tasknumber = 5
        n = input("Please enter your name (max. 10 characters): ")
        self.name = n[:10]

    def play(self):
        for i in range(1,self.tasknumber+1):
            a = Task(i, self.tasknumber)
            print(a)
            self.correct += a.answer()

    def measure(self, start):
        if start:
            self.starttime = time.time()
        else:
            self.time = round(time.time() - self.starttime, 2)

    def __str__(self):
        output = f"Correct: {self.correct} of" \
                    f" {self.tasknumber} in {self.time} seconds"
        if self.correct == self.tasknumber:
            output += ", Highscore"
            hs = Highscore()
            hs.save(self.name, self.time)
            print(hs)
        else:
            output += ", Sorry, no highscore"
        return output
```

Listing 8.30 game_file_oop.py File, Game Class

The random generator is initialized in the constructor of the class. The counter for correctly solved tasks is set to 0, the number of tasks to 5. The name of the player is determined. In addition, three properties of the Game class are set: correct, tasknumber, and name.

A total of five tasks are set and answered in the play() method. The measure() method is used to measure time. The time property of the Game class is set.

If all tasks are solved correctly, an object of the Highscore class is created and saved in the output method. Then, the highscore list is displayed.

The new Highscore class looks a little bit different.

```
class Highscore:
    def __init__(self):
        self.list = []
        if not glob.glob("highscore.csv"):
            return
        f = open("highscore.csv")
        line = f.readline()
        while line:
            split = line.split(";")
            name = split[0]
            time_required = split[1]
            self.list.append([name, float(time_required)])
            line = f.readline()
        f.close()

    def change(self, name, time_required):
        found = False
        for i in range(len(self.list)):
            if time_required < self.list[i][1]:
                self.list.insert(i, [name, time_required])
                found = True
                break
        if not found:
            self.list.append([name, time_required])

    def save(self, name, time_required):
        self.change(name, time_required)
        f = open("highscore.csv", "w")
        for element in self.list:
            name = element[0]
            time_required = element[1]
            f.write(f"{name};{time_required}\n")
        f.close()
```

```
def __str__(self):
    if not self.list:
        return "No highscores available"

    output = " P. Name              Time\n"
    for i in range(len(self.list)):
        output += f"{i+1:2d}. {self.list[i][0]:10}" \
                      f"{self.list[i][1]:6.2f} sec\n"
        if i >= 9:
            break
    return output
```

Listing 8.31 game_file_oop.py File, Highscore Class

The most important property is set in the constructor of the class: the highscore list named list. It is initially empty. If there is no CSV file, the list remains empty. However, if a CSV file exists, all lines are read from it. The lines are then split. The two components of a line are each added as a sub-list to the highscore list.

The change() method is required within the class by the save() method. A newly determined highscore is either inserted into or appended to the list.

The highscore list is first changed in the save() method. The entire list is saved in the CSV file and is also published in the output method if the file is not empty.

Chapter 9

Databases

A database is used to store large amounts of data, to process selected data, and to display data in a clear way. You can access databases using a database system and *Structured Query Language (SQL)*. In this book, we use two different SQL-based database systems:

- SQLite: This system is based on text files and is suitable for small amounts of data.
- MySQL: This complex database system is also suitable for large volumes of data.

9.1 Structure of Databases

Various tables can exist within a database. Table 9.1 shows a simple table with some personal data.

Name	First Name	Personnel Number	Salary	Birthday
Mayer	John	6714	3,500.00	3/15/1962
Smith	Peter	81343	3,750.00	4/12/1958
Mertins	Judith	2297	3,621.50	12/30/1959

Table 9.1 Example with Personal Data

The terms in the first row are referred to as the *fields* of the table. This row is followed by the individual *data records* of the table; in our example, we have three data records.

Nobody creates a database with a table for just three data records. However, this structure could also be used for several thousand data records. The fields each have a specific data type. In this case, we are using the *text, integer, number with decimal places,* and *date* data types.

You can create a database by following these steps:

1. Creating the database
2. Creating tables by specifying the structure
3. Entering the data records in the tables

You can also change the structure of a database after data has already been saved in it. However, this can easily lead to data loss, and it is therefore better to plan the structure thoroughly in advance.

We'll be using SQL statements to create the structure of databases and tables and to edit the data records (i.e., create, display, change, and delete).

9.2 SQLite

SQLite is a small and easy-to-use database system. Integrated directly into Python via the sqlite3 module, SQLite works on the basis of text files and can be used for smaller amounts of data. It does not require any additional installations.

The data types in SQLite are not as fixed as in other database systems; instead, they should be understood as recommendations. The following data types are available by default:

- TEXT: For character strings
- INTEGER: For integers
- REAL: For numbers with decimal places
- BLOB: For *binary large objects* (i.e., large binary datasets)
- NULL: Corresponds to None in Python

You must take care of checking and correctly converting values of other data types from SQLite to Python, for example, converting an SQLite string into a Python variable for a date.

> **Note**
> The data types we just listed and other SQL-specific information are written in capital letters, which is not necessary for SQL but does make things clearer.

9.2.1 Database, Table, and Data Records

In the next program, we'll first create an SQLite database and then a table with a unique index. After that, three data records will be created in the table. The structure and data from Table 9.1 are used for this purpose.

A unique index is used to identify individual data records. In the table containing the personal data, the personnelnumber field is suitable for this purpose. Let's look at this program now.

```
import os, sys, sqlite3
if os.path.exists("company.db"):
    print("Database already exists")
```

```
    sys.exit(0)

connection = sqlite3.connect("company.db")
cursor = connection.cursor()
sql = "CREATE TABLE people(" \
      "name TEXT, " \
      "firstname TEXT, " \
      "personnelnumber INTEGER PRIMARY KEY, " \
      "salary REAL, " \
      "birthday TEXT)"
cursor.execute(sql)

sql = "INSERT INTO people VALUES('Mayer', " \
      "'John', 6714, 3500, '3/15/1962')"
cursor.execute(sql)
connection.commit()
sql = "INSERT INTO people VALUES('Smith', " \
      "'Peter', 81343, 3750, '4/12/1958')"
cursor.execute(sql)
connection.commit()
sql = "INSERT INTO people VALUES('Mertins', " \
      "'Judith', 2297, 3621.5, '12/30/1959')"
cursor.execute(sql)
connection.commit()

connection.close()
```

Listing 9.1 sqlite_generate.py File

First, the system checks whether the database file *company.db* already exists.

If the file does exist, the program terminates with a corresponding message. Otherwise, the further course of the program would trigger an error. If you want to create a new database, you must first delete the database file *company.db*.

The connect() function from the sqlite3 module is used to establish a connection to the database. If the database does not yet exist, it will be created. The return value of the connect() function is an object of the sqlite3.connection class. The database can now be accessed via this connection.

The cursor() method is called for the connection object. It returns an object of the sqlite3.cursor class as the return value. This cursor object can be used to send SQL commands to the database and receive the results.

A string containing an SQL command is built. SQL commands are sent to the database using the execute() method of the cursor object.

The CREATE TABLE SQL statement creates a table in a database. After the name of the table (in our case, people), the names of the individual fields of the table with their respective data type are provided in parentheses. The type of the name, firstname, and birthday fields is TEXT. The type of the salary field is REAL, while the personnelnumber field is of the INTEGER type.

A primary key is defined in the personnelnumber field using the PRIMARY KEY addition. This index is a unique integer index for identifying data records. It ensures that no two data records can have the same personnel number.

The SQL statement INSERT INTO is used to insert data records. This action query leads to a change in the table. After the name of the table and the VALUES keyword, the individual values of the data record for the fields of the table are specified in parentheses. You must put strings in single quotation marks. In addition, the order of the fields must correspond to the order in which the table was created.

An action query should always be followed by a call of the commit() method of the connection object so that the changes are executed immediately.

Finally, the connection to the database is closed again using the close() method.

9.2.2 Displaying Data

All data records in the table are displayed with their entire contents in the next program.

```
import os, sys, sqlite3
if not os.path.exists("company.db"):
    print("No database")
    sys.exit(0)


connection = sqlite3.connect("company.db")
cursor = connection.cursor()
sql = "SELECT * FROM people"
cursor.execute(sql)


for dr in cursor:
    print(dr[0], dr[1], dr[2], dr[3], dr[4])
connection.close()
```

Listing 9.2 sqlite_display.py File

The program generates the following output:

```
Mertins Judith 2297 3621.5 12/30/1959
Mayer John 6714 3500.0 3/15/1962
Smith Peter 81343 3750.0 4/12/1958
```

If the database file does not exist, the program will terminate.

After creating the database connection and the data record cursor, an SQL query is sent using the execute() method.

The SELECT SQL statement is used to formulate selection queries to select database contents. The SELECT * FROM people statement returns all data records with all fields of the table. The asterisk (*) represents "all fields".

After running the execute() method, the result of the query is available in the cursor object. These are the data records that match the query. The sequence of these data records can be run through using a loop. Each data record corresponds to a tuple consisting of the values for the individual fields in the same order as when the table was created.

Since SELECT is not an action query (which makes a change in the database), but only a selection query (which determines information), the commit() method does not need to be called here.

Finally, the connection to the database is terminated again.

9.2.3 Selecting Data

You can use SELECT and WHERE to filter out one or more data records. Here are some examples:

```
import os, sys, sqlite3
if not os.path.exists("company.db"):
    print("No database")
    sys.exit(0)

connection = sqlite3.connect("company.db")
cursor = connection.cursor()

cursor.execute("SELECT name, firstname FROM people")
for dr in cursor:
    print(dr[0], dr[1])
print()

cursor.execute("SELECT * FROM people WHERE salary > 3600")
for dr in cursor:
    print(dr[0], dr[3])
print()

cursor.execute("SELECT * FROM people WHERE name = 'Smith'")
for dr in cursor:
    print(dr[0], dr[1])
```

```
print()

cursor.execute("SELECT * FROM people WHERE salary >= 3600 AND salary <= 3650")
for dr in cursor:
    print(dr[0], dr[3])

connection.close()
```

Listing 9.3 sqlite_select.py File

The following output is generated:

```
Mertins Judith
Mayer John
Smith Peter

Mertins 3621.5
Smith 3750.0

Smith Peter

Mertins 3621.5
```

Four independent SQL queries are executed.

If you only want to see the values of specific fields, you need to write the names of these fields between SELECT and FROM. The SELECT name, firstname FROM people statement only returns the contents of the two fields mentioned for all data records in the table.

You can use WHERE to restrict the selection of data records by creating a condition with comparison operators.

The SELECT * FROM people WHERE salary > 3600 statement returns the contents of all fields for the data records for which the value of the salary field is above the specified limit (see also Table 9.2).

Operator	Explanation
=	Equal to
<>	Unequal to
>	Greater than
>=	Greater than or equal to
<	Less than
<=	Less than or equal to

Table 9.2 Comparison Operators in SQL

You can also compare strings but pay attention to the single quotation marks. The SELECT * FROM people WHERE name = 'Smith' SQL statement only returns the data records for which the name is the same as "Smith".

Logical operators make it possible to link multiple conditions. The SELECT * FROM people WHERE salary >= 3600 AND salary <= 3650 statement only returns the data records for which the value of the field salary is within the specified limits (see also Table 9.3).

Operator	Explanation
NOT	Logical NOT: The truth value of a condition is reversed.
AND	Logical AND: Both conditions must be met.
OR	Logical OR: Only one of the conditions must be met.

Table 9.3 Logical Operators in SQL

9.2.4 The LIKE Operator

You can use the LIKE operator to compare character strings. This operator is particularly useful when selecting data records. Placeholders are also allowed in this context. You can therefore create queries such as the following:

- Search all data records beginning with …
- Search all data records ending with …
- Search all data records that contain …

Let's look at some examples next.

```
import os, sys, sqlite3
if not os.path.exists("company.db"):
    print("No database")
    sys.exit(0)


connection = sqlite3.connect("company.db")
cursor = connection.cursor()

cursor.execute("SELECT * FROM people WHERE name LIKE 'm%'")
for dr in cursor:
    print(dr[0], dr[1])
print()

cursor.execute("SELECT * FROM people WHERE name LIKE '%i%'")
for dr in cursor:
    print(dr[0], dr[1])
print()
```

```
cursor.execute("SELECT * FROM people WHERE name LIKE 'M_er'")
for dr in cursor:
    print(dr[0], dr[1])

connection.close()
```

Listing 9.4 sqlite_like.py File

The following output is generated:

```
Mertins Judith
Mayer John

Mertins Judith
Smith Peter

Mayer John
```

Three independent SQL queries are executed.

The % placeholder (percent sign) represents an indefinite number of characters. The SELECT * FROM people WHERE name LIKE 'm%' statement only returns the data records in which the name begins with the letter "m" (regardless of uppercase and lowercase).

If the percent sign appears before *and* after a certain character, all data records containing this character are searched for. The SELECT * FROM people WHERE name LIKE '%i%' statement only returns the data records in which the letter "i" (or "I") occurs.

The _ placeholder (underscore) represents any single character. The SELECT * FROM people WHERE name LIKE 'M_er' statement only returns the data records in which the name begins with "M" and ends with "er" after two further characters. This applies, for example, to "Maier", "Mayer", "Meier", and "Meyer".

9.2.5   Sorting the Output

ORDER BY provides selected data records in a sorted order. You can sort this list of records by one or more fields, in ascending or descending order. Let's consider some examples.

```
import os, sys, sqlite3
if not os.path.exists("company.db"):
    print("No database")
    sys.exit(0)

connection = sqlite3.connect("company.db")
cursor = connection.cursor()

cursor.execute("SELECT * FROM people ORDER BY salary DESC")
```

```
for dr in cursor:
    print(dr[0], dr[1], dr[3])
print()

cursor.execute("SELECT * FROM people ORDER BY name, firstname")
for dr in cursor:
    print(dr[0], dr[1])

connection.close()
```

Listing 9.5 sqlite_sort.py File

Now, let's assume that another data record was previously added for this query for a person called "Patrick Mayer" with a salary of $3,810. The output then reads as follows:

```
Mayer Patrick 3810.0
Smith Peter 3750.0
Mertins Judith 3621.5
Mayer John 3500.0

Mayer John
Mayer Patrick
Mertins Judith
Smith Peter
```

Two independent SQL queries are executed.

The SELECT * FROM people ORDER BY salary DESC statement returns all data records, sorted by salary in descending order. The DESC addition stands for *descending*. To use ascending sort order, you could use the ASC addition. However, since ascending order is the default sorting option, you can simply omit ASC.

The SELECT * FROM people ORDER BY name, firstname statement returns all data records, sorted by name in ascending order. If any (family) names are identical, the sort order will be based on the first name for those records.

9.2.6 Selection after Entry

Data records can also be selected after an entry has been made. The next program only displays the data records that correspond to specific entries.

```
import os, sys, sqlite3
if not os.path.exists("company.db"):
    print("No database")
    sys.exit(0)
```

```
connection = sqlite3.connect("company.db")
cursor = connection.cursor()

entryName = input("Please enter the family name you are looking for: ")
entryFirstName = input("Please enter the first name you are looking for: ")
sql = "SELECT * FROM people WHERE name LIKE ? OR firstname LIKE ?"
cursor.execute(sql, (entryName, entryFirstName))
for dr in cursor:
    print(dr[0], dr[1])
print()

entry = input("Please enter the part of the name you are looking for: ")
sql = "SELECT * FROM people WHERE name LIKE ?"
entry = "%" + entry + "%"
cursor.execute(sql, (entry,))
for dr in cursor:
    print(dr[0], dr[1])

connection.close()
```

Listing 9.6 sqlite_input.py File

The output (with a sample input) looks as follows:

```
Please enter the family name you are looking for: Mayer
Please enter the first name you are looking for: Peter
Mayer John
Smith Peter

Please enter the part of the name you are looking for: r
Mertins Judith
Mayer John
```

In the first part of the program, "Mayer" and "Peter" are entered. An SQL command is sent that uses these two entries. The parameter substitution of the Python database interface is used to prevent possible infiltration by malicious SQL code (also known as *SQL injection*).

For each question mark in the SQL code, an element of the tuple is used, which is specified as the second, optional parameter of the execute() method. There are two question marks in our example, so the tuple contains two elements. All data records in which the family name is "Mayer" or the first name is "Peter" are returned.

The "r" character is entered in the second part of the program. This entry is placed between the percentage signs which serve as placeholders. All data records whose name contains an "r" are provided.

There is only one question mark in the SQL command. The tuple therefore only contains one element. In such a case, a comma must be appended to the tuple; otherwise, an error will occur.

9.2.7   Changing Data Records

The UPDATE statement is used to change the content of one or more data records. Its structure is similar to that of the SELECT statement. Make sure you choose your selection criteria carefully so that you do not inadvertently make the wrong changes. Let's look at an example.

```
sql = "UPDATE people SET salary = 3800"
```

This statement would set the value for the salary field to the value 3800 for all data records in the people table—this action is certainly not intended.

For this reason, you should make a restriction, preferably via the field with the unique index, in this case, the personnelnumber field:

```
sql = "UPDATE people SET salary = 3780 WHERE personnelnumber = 81343"
```

Let's put this concept to work with a program that displays output before the change and an output after the change:

```
import os, sys, sqlite3
if not os.path.exists("company.db"):
    print("No database")
    sys.exit(0)

def output():
    sql = "SELECT * FROM people"
    cursor.execute(sql)
    for dr in cursor:
        print(dr[0], dr[1], dr[2], dr[3])
    print()

connection = sqlite3.connect("company.db")
cursor = connection.cursor()
output()

sql = "UPDATE people SET salary = 3780 WHERE personnelnumber = 81343"
cursor.execute(sql)
connection.commit()
```

```
output()
connection.close()
```

Listing 9.7 sqlite_change.py File

The following output is generated:

```
Mertins Judith 2297 3621.5
Mayer John 6714 3500.0
Smith Peter 81343 3750.0

Mertins Judith 2297 3621.5
Mayer John 6714 3500.0
Smith Peter 81343 3780.0
```

Notice how only one data record has been changed. The following example changes the contents of several fields of a selected data record:

```
sql = "UPDATE people SET salary = 3780, firstname = 'John-Peter' " \
      "WHERE personnelnumber = 81343"
```

No change may be made that results in two or more data records having the same content in the field on which the unique index is located, in this case, the personnelnumber field.

Thus, the following statement ...

```
sql = "UPDATE people SET personnelnumber = 6714 " \
      "WHERE personnelnumber = 81343"
```

... would lead to the following error message:

```
sqlite3.IntegrityError: UNIQUE constraint failed: people.personnelnumber
```

The entries in the field with the PRIMARY KEY must be unique.

9.2.8   Deleting Data Records

The DELETE SQL statement enables you to delete data records. Make sure to choose your selection criteria carefully to avoid accidentally deleting the wrong data records or deleting all data records. Let's look at an example.

```
import os, sys, sqlite3
if not os.path.exists("company.db"):
    print("No database")
    sys.exit(0)

def output():
```

```
    sql = "SELECT * FROM people"
    cursor.execute(sql)
    for dr in cursor:
        print(dr[0], dr[1], dr[2], dr[3])
    print()

connection = sqlite3.connect("company.db")
cursor = connection.cursor()
output()

sql = "DELETE FROM people WHERE personnelnumber = 8339"
cursor.execute(sql)
connection.commit()

output()
connection.close()
```

Listing 9.8 sqlite_delete.py File

Let's assume that another data record was previously added for this query for "Patrick Mayer" with the personnel number 8339 and a salary of $3,810. The output reads:

```
Mertins Judith 2297 3621.5
Mayer John 6714 3500.0
Mayer Patrick 8339 3810.0
Smith Peter 81343 3780.0

Mertins Judith 2297 3621.5
Mayer John 6714 3500.0
Smith Peter 81343 3780.0
```

The data record with personnel number 8339 has been deleted. You can recognize this deletion has occurred by comparing the outputs before and after the deletion.

9.2.9   Exercise

u_sqlite_language

Now, let's create a simple language learning program based on an SQLite database in the *u_sqlite_language.py* file. Let's start with how the program should work.

```
Your selection: 0 (= End), 1 (= All), 2 (= Learn), 3 (= Test): 1
1 / Bedingung / condition
2 / suchen / to look for
3 / Werbeanzeige / advertisement
```

```
4 / abkürzen / to abbreviate
5 / nützlich / useful
6 / Wirkung / effect
7 / beraten / to advise
8 / übersetzen / to translate
9 / simple / easy
10 / ankündigen / to announce
Your selection: 0 (= End), 1 (= All), 2 (= Learn), 3 (= Test): 4
Your selection: 0 (= End), 1 (= All), 2 (= Learn), 3 (= Test): 2
ankündigen / to announce
Your selection: 0 (= End), 1 (= All), 2 (= Learn), 3 (= Test): 2
beraten / to advise
Your selection: 0 (= End), 1 (= All), 2 (= Learn), 3 (= Test): 2
Bedingung / condition
Your selection: 0 (= End), 1 (= All), 2 (= Learn), 3 (= Test): 3
Please translate: 'Bedingung' ==> condition
Correct, word is removed from the test
Please translate: 'nützlich' ==> useful
Correct, word is removed from the test
Please translate: 'abkürzen' ==> to abbreviate
Correct, word is removed from the test
Please translate: 'Werbeanzeige' ==> advertment
Unfortunately wrong, correct would be: advertisement
Please translate: 'einfach' ==> easy
Correct, word is removed from the test
Please translate: 'Wirkung' ==> effect
Correct, word is removed from the test
Please translate: 'ankündigen' ==> to announce
Correct, word is removed from the test
Please translate: 'beraten' ==> to advise
Correct, word is removed from the test
Please translate: 'suchen' ==> to look for
Correct, word is removed from the test
Please translate: 'Werbeanzeige' ==> advertisement
Correct, word is removed from the test
Please translate: 'übersetzen' ==> to translate
Correct, word is removed from the test
Test completed successfully
Your selection: 0 (= End), 1 (= All), 2 (= Learn), 3 (= Test): 0
```

After you have launched the program, a menu appears. You can enter only values from 0 to 3 to select a menu item. If a different value is entered, the menu appears again. You can choose from the following options:

- 0: The program is terminated.
- 1: To serve as an overview, all ten words are displayed in both German and English and with consecutive numbers.
- 2: A randomly selected word pair is displayed for learning.
- 3: A test begins. Randomly selected German words are displayed one after the other. The corresponding English word must be entered in each case. If the entry is correct, the word pair is removed from the list. If the entry is wrong, the correct solution is displayed, and the word is asked again later. If all the words have been translated correctly, the list is empty, and the test ends.

If you're not yet familiar with the development of the program, you can find some tips in Appendix A, Section A.5.

Note

In Chapter 11, Section 11.3, a graphical user interface (GUI) is created using *PyQt*. PyQt is an interface between Python and the widely used *Qt* library. The GUI is used to manage the SQLite database *company.db*. This comprehensive application combines two important technologies and other elements of Python programming.

9.3 MySQL

MySQL is a complex database management system. In this section, I'll introduce you to MySQL through some simple programs. With regard to the examples in this book, whether MySQL or its spin-off MariaDB is used makes no difference.

You're already familiar with the various SQL commands for selecting or changing data from Section 9.2. For this reason, in this section, only a database with one table and a few data records is created and queried in a simple way.

MySQL provides a whole range of data types, including the following:

- VARCHAR for character strings
- INT for integers
- DOUBLE for numbers with decimal places
- DATE for dates

9.3.1 XAMPP and Connector/Python

A local MySQL database server is required to test the programs. This local server can be found in the *XAMPP* package, which is preconfigured and easy to install. The installation of XAMPP and starting the server is described in Appendix A, Section A.3.

The *Connector/Python* driver provides an interface between Python and MySQL. You can install it with the following commands (see also Appendix A, Section A.1):

- On Windows and macOS: `pip install mysql-connector-python`
- On Ubuntu Linux: `sudo apt install python3-mysql.connector`

Note the different name used in the Ubuntu Linux command.

9.3.2   Database, Table, and Data Records

Similar to SQLite, as described in Section 9.2.1, our program creates a MySQL database first and then a table with a unique index. Finally, three data records will be created in the table.

```
import sys, mysql.connector
try:
    connection = mysql.connector.connect \
        (host = "localhost", user = "root", passwd = "")
except:
    print("No connection to the server")
    sys.exit(0)

cursor = connection.cursor()

cursor.execute("CREATE DATABASE IF NOT EXISTS company")
connection.commit()

cursor.execute("USE company")
connection.commit()

cursor.execute("CREATE TABLE IF NOT EXISTS people"
        "(name VARCHAR(30), firstname VARCHAR(25),"
        "personnelnumber INT(11), salary DOUBLE, "
        "birthday DATE, PRIMARY KEY (personnelnumber))"
        "ENGINE=InnoDB DEFAULT CHARSET=UTF8")
connection.commit()

try:
    cursor.execute("INSERT INTO people VALUES('Mayer', " \
        "'John', 6714, 3500, '1962-03-15')")
    connection.commit()
    cursor.execute("INSERT INTO people VALUES('Smith', " \
        "'Peter', 81343, 3750, '1958-04-12')")
    connection.commit()
    cursor.execute("INSERT INTO people VALUES('Mertins', " \
```

```
        "'Judith', 2297, 3621.5, '1959-12-30')")
    connection.commit()
except:
    print("Error during creation")

cursor.close()
connection.close()
```

Listing 9.9 mysql_generate.py File

After importing the mysql.connector module, a connection to the database server is established using the connect() function. The name of the server (in this case, local-host); the name of the user (in this case, root); and the password (in this case, no password) are specified. If the MySQL database server is not running, the connection will fail. The return value of the connect() function is a connection object. The database can be accessed via this connection.

The cursor() method is then used to create a cursor object. This cursor can be used to send SQL commands to the database and receive the results.

The execute() method is used to send SQL commands. The SQL command CREATE DATA-BASE company creates a new database named company. The IF NOT EXISTS addition ensures that the database is only created if it does not exist yet.

Calling the commit() method of the connection object causes the SQL statement to be executed immediately.

The USE company SQL command is used to select the company database after it has been created.

The people table is created with five fields. The name and firstname fields are of type VARCHAR. A maximum length of 30 and 25 characters is selected. The personnelnumber field is of type INT. The size 11 is the default value for the type INT. The salary field is of type DOUBLE for numbers with decimal places. The birthday field is of type DATE for dates. The IF NOT EXISTS addition also means that the table will only get created if it does not exist yet.

A primary key is defined in the personnelnumber field. This key ensures that no two entries can have the same personnel number.

The INSERT INTO SQL command is used to create three data records. The data is entered in the order of the fields. Strings are enclosed in single quotation marks. Decimal places are separated by a decimal point. A date is entered in the format YYYY-MM-DD and also within single quotation marks.

A second call of the same program would lead to an error message, as a primary key is defined in the personnelnumber field. No entry can be created that has the same personnel number as an existing data record.

At the end of these actions, the `cursor` object and the `connection` object are closed again using the `close()` method.

The program has no output. XAMPP contains the *phpMyAdmin* program for managing the MySQL server databases. You can use this program to verify the existence of the new database, the table, and the data records. Alternatively, you can call the program, which we'll describe next.

9.3.3   Displaying Data

Our next program displays all the data records in the table with all their contents.

```
import sys, mysql.connector
try:
    connection = mysql.connector.connect(host = "localhost", \
        user = "root", passwd = "", db = "company")
except:
    print("No connection to the server or no database")
    sys.exit(0)

cursor = connection.cursor()
cursor.execute("SELECT * FROM people")
result = cursor.fetchall()
cursor.close()
connection.close()

for dr in result:
    for field in dr:
        print(f"{field} ", end="")
    print()
```

Listing 9.10 mysql_display.py File

The program generates the following output:

```
Mertins Judith 2297 3621.5 1959-12-30
Mayer John 6714 3500.0 1962-03-15
Smith Peter 81343 3750.0 1958-04-12
```

The `connect()` function is called with a fourth parameter. Not only does this function establish a connection to the database server but also to the database on the server.

All data records are queried using the `SELECT` SQL command. As this command is not an action query, only a selection query, the `commit()` method does not need to be called at this point.

The `fetchall()` method of the `cursor` object returns a list containing the result of the query. The individual data records, with the values of their fields, are output.

9.4 Our Game: Version with Highscore Database

We now want to make another version of our mental calculation game. The player's name and time needed are stored permanently in an SQLite database so that a high-score list can be maintained.

The difference here, as compared to the version with the highscore file in Chapter 8, Section 8.10, is that no separate list needs to be maintained. Instead, the new values are written directly "unsorted" to the database. Later, the existing values are read directly from the database in sorted order and displayed on the screen.

This approach has several advantages:

- You don't require any exchange between the list and the file.
- The sorting is carried out by SQL and does not need to take place within the Python program.
- If the program is terminated prematurely, no data is lost because the data is saved after each game.

Let's look at the first part of this program.

```python
import random, time, glob, sqlite3

def hs_show():
    if not glob.glob("highscore.db"):
        print("No highscores available")
        Return

    con = sqlite3.connect("highscore.db")
    cursor = con.cursor()
    sql = "SELECT * FROM data ORDER BY time LIMIT 10"
    cursor.execute(sql)

    print(" P. Name            Time")
    i = 1
    for dr in cursor:
        print(f"{i:2d}. {dr[0]:10} {dr[1]:5.2f} sec")
        i = i+1
    con.close()
```

Listing 9.11 game_sqlite.py File, Part 1 of 3

The hs_show() function determines whether the database file exists. If it does, a connection is established.

The SELECT statement is used to select the first ten data records, sorted in ascending order by playing time. The LIMIT addition in the SQL query limits the number. Once the data records have been output, the connection is terminated again.

Let's now turn to the second part of this program.

```
def game():
    ...
    if not glob.glob("highscore.db"):
        con = sqlite3.connect("highscore.db")
        cursor = con.cursor()
        sql = "CREATE TABLE data(name TEXT, time REAL)"
        cursor.execute(sql)
        con.close()

    con = sqlite3.connect("highscore.db")
    cursor = con.cursor()
    sql = f"INSERT INTO data VALUES('{name}', {difference})"
    cursor.execute(sql)
    con.commit()
    con.close()

    hs_show()
```

Listing 9.12 game_sqlite.py File, Part 2 of 3

The beginning of the game() function does not differ from that of the version with the highscore file (see Chapter 8, Section 8.10). The changes do not become apparent until the data is saved.

The program first determines whether the database file exists. If not, the database will be created with the table.

The newly determined values are written to the database as a new data record. Then, the highscore list is displayed.

Finally, let's look at our main program.

```
random.seed()

while True:
    try:
        menu = int(input("Please enter "
            "(0: End, 1: Highscores, 2: Play): "))
```

```
except:
    print("Wrong input")
    continue

if menu == 0:
    break
elif menu == 1:
    hs_show()
elif menu == 2:
    game()
else:
    print("Wrong input")
```

Listing 9.13 game_sqlite.py File, Part 3 of 3

Compared to the version with a highscore file (see Chapter 8, Section 8.10), the calls of
the two functions for exchanging data between the list and the file are missing. An endless loop runs in the main program. The highscores can be displayed, the game can be
played, or the program can be closed.

9.5   Our Game: Object-Oriented Version with Highscore Database

This version of the mental calculation game is based on the object-oriented version
with a highscore file from Chapter 8, Section 8.11. Only the Highscore class has been
changed so that the advantages of a database can be used.

```
import random, time, glob, sqlite3
...
class Highscore:
    def save(self, name, time):
        if not glob.glob("highscore.db"):
            con = sqlite3.connect("highscore.db")
            cursor = con.cursor()
            sql = "CREATE TABLE data(name TEXT, time REAL)"
            cursor.execute(sql)
            con.close()

        con = sqlite3.connect("highscore.db")
        cursor = con.cursor()
        sql = f"INSERT INTO data VALUES('{name}', {time})"
        cursor.execute(sql)
        con.commit()
        con.close()
```

```
def __str__(self):
    if not glob.glob("highscore.db"):
        return "No highscores available"

    con = sqlite3.connect("highscore.db")
    cursor = con.cursor()
    sql = "SELECT * FROM data ORDER BY time LIMIT 10"
    cursor.execute(sql)

    output = " P. Name                Time\n"
    i = 1
    for dr in cursor:
        output += f"{i:2d}. {dr[0]:10} {dr[1]:5.2f} sec\n"
        i = i+1

    con.close()
    return output
```

Listing 9.14 game_sqlite_oop.py File

The Highscore class contains only the two methods for saving and outputting the highscore.

The save() method checks whether a database file exists. If the file does not exist, the database and the table will be created. The newly determined data is added.

In the output method, the data records are selected in sorted order and then output with formatting.

Chapter 10
User Interfaces

To operate the programs described so far in this book, we've used entries in the Integrated Development and Learning Environment (IDLE) or via the command line. However, to create convenient *graphical user interfaces (GUIs)*, we'll need to turn to the *Tk* and *Qt* libraries. Both libraries can be used rather uniformly on any operating system. When you work with GUIs, event-oriented programming comes into play.

The *tkinter* module is an *interface* to the *Tk* library. This interface is immediately available once you have installed Python.

The *PyQt6* module provides an interface to the *Qt* library. This interface must be installed in an additional step.

10.1 Introduction

When you use the *tkinter* interface, the creation of a GUI application comprises the following steps:

1. Creation of a window with the elements of the user interface.
2. Creation and arrangement of control elements (in this context, called *widgets*) within the window for operating the application.
3. Start of an endless loop in which the operation of the widgets leads to events and thus to the execution of the program's components.

10.1.1 Our First GUI Program

Our first program serves as an introduction to GUI programming.

```
import tkinter

def end():
    window.destroy()

window = tkinter.Tk()
window.resizable(0, 0)

buEnd = tkinter.Button(window, text="End", command=end, width=10)
```

```
buEnd.pack(padx=30, pady=30)
```

```
window.mainloop()
```

Listing 10.1 gui_first.py File

The GUI is visually adapted to the operating system on which it is running and thus has a familiar design.

The results for Windows, Ubuntu Linux, and macOS are shown in Figure 10.1 through Figure 10.3. The rest of the screenshots in this chapter are taken from the user interface on Windows.

When you click the **End** button, the application will be closed.

Figure 10.1 Our First Program on Windows

Figure 10.2 Our First Program on Ubuntu Linux

Figure 10.3 Our First Program on macOS

The imported tkinter module and its classes are required to create the interface elements. The end() function is defined, in which the application window is closed using the destroy() method and thus the entire application is ended.

An object of the Tk class is created. This object corresponds to a window—in this case, the application window, which contains all the widgets of the application. The constructor of the Tk class returns a reference to the Tk object, which is stored in the window variable. You can use this variable to access the application window.

Call the resizable() method with the values 0, 0 to specify that the size of the applica-tion window cannot be changed.

An object of the Button class is created. This widget serves as a button for triggering actions. The properties of a widget are defined using the parameters of the constructor. Many properties of a widget can be changed later.

The first parameter specifies in which surrounding element the button is embedded—in this case, in the application window.

The named text parameter defines the name of the button. The named command param-eter determines the name of the callback function, which is executed when the button is clicked (see also Chapter 5, Section 5.6.9). This is the end() function. You can use the width parameter to specify the width of the button. The constructor of the Button class returns a reference to the Button object, which is stored in the buEnd variable. You can access the button via this variable.

A widget does not appear in the application window until it has been arranged using a geometry manager. Calling the pack() method calls the *pack* geometry manager. It arranges widgets one after the other in the surrounding element in a simple way.

You can use the padx and pady parameters to create a distance in the x-direction or y-direction between the widget and the edge of the surrounding element.

When you use the *pack* geometry manager, the size of the application window depends on the widgets it contains, while also taking into account the padding.

At the end, the mainloop() method calls an endless loop for the application window. The loop ensures that the window with the widgets is permanently displayed.

> **Note**
>
> As with other Python applications, you can start the GUI applications either from IDLE or from the command line.

10.1.2 Setup and Arrangement

Our next program sets up a GUI with several widgets. You'll learn about the display and arrangement of widgets in the application window through this example.

```python
import tkinter

def hello():
    lbOutput["text"] = "Hello"

def end():
    window.destroy()
```

10   User Interfaces

```
window = tkinter.Tk()
window.title("GUI")
window.resizable(0, 0)

buHello = tkinter.Button(window, text="Hello", command=hello, width=10)
buHello.grid(row=0, column=0, sticky="w", padx=5, pady=5)

lbOutput = tkinter.Label(window, text="(empty)",
    anchor="w", relief="sunken", width=20)
lbOutput.grid(row=1, column=0, padx=5, pady=5)

buEnd = tkinter.Button(window, text="End", command=end, width=10)
buEnd.grid(row=2, column=1, padx=5, pady=5)

window.mainloop()
```

Listing 10.2 gui_arrangement.py File

The output of the program is shown in Figure 10.4.

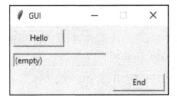

Figure 10.4 Arrangement of Widgets

The application window contains three widgets, namely, two buttons and a label widget (which is an object of the Label class). This label is used to output a text or display an image and can be accessed here using the lbOutput variable. The text and width parameters have the same meaning as for a button.

If the **Hello** button is clicked, the "Hello" text displays in the label below. For this purpose, a new value is assigned to the text property of the label in the hello() function. The lbOutput variable serves as a reference to a dictionary for the properties of the widget. The property and its values form the key-value pairs.

The title() method is called for the application window. This method is used to assign the title that can be seen in the header of the window.

The grid() method is called for the three widgets to use the *grid* geometry manager with its many possibilities. The widgets are displayed in a *grid* of cells, similar to the cells in a table. The row and column parameters enable you to determine the position of the cell within the grid. The numbering starts at 0.

If the surrounding element (in our case, the cell of the grid) is larger than a widget, you can use the sticky parameter to specify the edge of the surrounding element to which the widget is oriented.

The possible values for the sticky parameter are:

- w: To the western edge (i.e., the left edge)
- e: To the eastern edge (i.e., the right edge)
- n: To the northern edge (i.e., the top edge)
- s: To the southern edge (i.e., the bottom edge)

You can also form combinations of these values. If sticky is not used, the widget is placed in the center of the cell.

In this example, the label is wider than the button and therefore determines the width of the column on the left. The button is positioned on the left edge of the cell using the sticky parameter.

You can use the relief parameter to influence the display of a widget. How a label is displayed can be changed using the following possible values:

- flat: Flat, without any recognizable edge
- groove: With a thin sunken frame
- ridge: With a thin raised frame
- raised: Raised as a whole, like a button
- sunken: Deepened as a whole
- solid: With a solid black frame

If a widget is larger than the text displayed in it, you can use the anchor parameter to specify on which side of the widget the text is anchored (i.e., how the text is aligned). The possible values are the same as for the sticky parameter (i.e., w, e, n, s, or a combination of these values). If anchor is not used, the text is placed in the center of the widget.

In this example, the label widget is wider than the text, which is arranged on the left edge of the label using the anchor parameter.

The size of the application window depends on the widgets it contains, taking into account the padding.

10.1.3 Event Methods for Multiple Widgets

By default, you can only specify the name of a callback function for the response to a widget event; you cannot specify parameters.

To remedy this, a lambda function (also known as an anonymous function) can be used; see Chapter 5, Section 5.6.8. In the following example, two buttons call the same function but with different parameters.

```
import tkinter

def out(p):
    lbOutput["text"] = p

def end():
    window.destroy()

window = tkinter.Tk()
window.title("GUI")
window.resizable(0, 0)

buOne = tkinter.Button(window, text="1", width=10, command=lambda:out(1))
buOne.grid(row=0, column=0, padx=5, pady=5)

buTwo = tkinter.Button(window, text="2", width=10, command=lambda:out(2))
buTwo.grid(row=1, column=0, padx=5, pady=5)

lbOutput = tkinter.Label(window, width=20)
lbOutput.grid(row=2, column=0, padx=5, pady=5)

window.mainloop()
```

Listing 10.3 gui_lambda.py File

When you click one of the two buttons, an anonymous function will be called. In the only statement of this function, the out() function is called with the parameter 1 or 2. The value of the parameter is then displayed in the label, as shown in Figure 10.5.

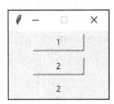

Figure 10.5 After Clicking the 2 Button

10.2   Widget Types

This section describes other widgets: entry, text, listbox, spinbox, radiobutton, checkbutton, and scale.

10.2.1   Single-Line Input Field

An entry widget is a single-line input field for entering short texts or numbers. Our next example illustrates the processing of data that has been entered. After entering a number in the entry widget, this value is doubled and then output.

```python
import tkinter

def double():
    try:
        number = float(etInput.get())
        lbOutput["text"] = number * 2
    except:
        lbOutput["text"] = "No number"

def end():
    window.destroy()

window = tkinter.Tk()
window.title("Single-line")
window.resizable(0, 0)

lbInput = tkinter.Label(window, text="Your input:")
lbInput.grid(row=0, column=0, sticky="w", padx=5, pady=5)

etInput = tkinter.Entry(window)
etInput.grid(row=1, column=0, padx=5, pady=5)

buDouble = tkinter.Button(window, text="Double",
    command=double, width=10)
buDouble.grid(row=2, column=0, sticky="w", padx=5, pady=5)

lbOutput = tkinter.Label(window, text="(empty)")
lbOutput.grid(row=3, column=0, sticky="w", padx=5, pady=5)

buEnd = tkinter.Button(window, text="End", command=end, width=10)
buEnd.grid(row=4, column=1, padx=5, pady=5)

window.mainloop()
```

Listing 10.4 gui_input.py File

After entering a number and clicking the **Double** button, the result shown in Figure 10.6 is displayed.

10

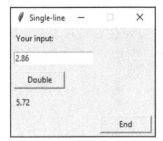

Figure 10.6 A Single-Line Input Field

No width is specified for the two labels, and therefore, they are only as wide as their contents. The width of the left column is determined by the width of the input field. The two labels and the button are arranged on the left edge of their cell using the sticky parameter.

A single-line input field is an object of the Entry class. A reference to this object is assigned to the etInput variable.

When the **Double** button is clicked, the double() function is called. The get() method returns a character string containing the entered text. If the text cannot be converted into a number using the float() function, an exception occurs, and an error message appears. Otherwise, the value of the number is doubled and output.

10.2.2   Hidden Input and Deactivating a Widget

An entry widget can also support the input of hidden information, such as passwords. In this case, the show parameter defines which character is displayed in an entry widget during input, as shown in Figure 10.7.

If the correct password is entered in the following program, the **Access permitted** message appears; otherwise, the **Access not permitted** message.

You can simplify the operation of a program by activating or deactivating a widget depending on the program status. In the next program, the **End** button is initially deactivated. The button does not become activated until the password has been checked.

```
import tkinter

def check():
    pw = etPassword.get()
    if pw == "Bingo":
        lbOutput["text"] = "Access permitted"
    else:
        lbOutput["text"] = "Access not permitted"
    etPassword.delete(0, "end")
    buEnd["state"] = "normal"
```

```
def end():
    window.destroy()

window = tkinter.Tk()
window.title("Password")
window.resizable(0, 0)

lbPassword = tkinter.Label(window, text="Your password:")
lbPassword.grid(row=0, column=0, sticky="w", padx=5, pady=5)

etPassword = tkinter.Entry(window, show="*")
etPassword.grid(row=1, column=0, padx=5, pady=5)

buCheck = tkinter.Button(window, text="Check",
    command=check, width=10)
buCheck.grid(row=2, column=0, sticky="w", padx=5, pady=5)

lbOutput = tkinter.Label(window, text="(empty)")
lbOutput.grid(row=3, column=0, sticky="w", padx=5, pady=5)

buEnd = tkinter.Button(window, text="End", command=end,
    width=10, state="disabled")
buEnd.grid(row=4, column=1, padx=5, pady=5)

window.mainloop()
```

Listing 10.5 gui_password.py File

The show parameter of the entry widget is assigned the value *, which is why an asterisk is displayed for each character entered, as is usual for passwords.

When the **Check** button is clicked, the check() function is called. The get() method returns the input. The system checks whether this corresponds to the correct password.

Figure 10.7 Hidden Input, Deactivated Widget

You can use the delete() method to delete a character string in an entry widget. The range of the character string to be deleted starts with the index in the first parameter and ends before the index in the second parameter. If the "end" value is specified as the second parameter, everything up to and including the last character will be deleted.

The button is created with the disabled value for the state parameter. It is therefore deactivated and cannot be operated. If you assign the default value normal to the state property, the button will activate.

10.2.3   Multiline Input Field

A text widget is a multiline input field that is used to enter and display large amounts of text. Compared to the entry widget, it provides more options for processing data.

The tkinter.scrolledtext module provides the ScrolledText widget, which is a further development of a simple text widget. This advanced widget enables a user to vertically scroll within the multiline input field with a scroll bar.

In the next program, the content of a text file is loaded into a ScrolledText widget. The contents of a ScrolledText widget can also be saved in a text file.

```python
import tkinter, tkinter.scrolledtext

def load():
    txText.delete(1.0, "end")
    try:
        f = open("gui_multiline.txt")
        txText.insert(1.0, f.read())
        f.close()
    except:
        txText.insert(1.0, "File not opened")

def save():
    try:
        f = open("gui_multiline.txt", "w")
        f.write(txText.get(1.0, "end"))
        f.close()
    except:
        txText.insert(1.0, "File not opened")

def delete():
    txText.delete(1.0, "end")

def end():
    window.destroy()
```

```
window = tkinter.Tk()
window.title("Multiline")
window.resizable(0, 0)

lbText = tkinter.Label(window, text="Text:")
lbText.grid(row=0, column=0, sticky="w", padx=5, pady=5)

txText = tkinter.scrolledtext.ScrolledText(window, width=50, height=4)
txText.grid(row=1, column=0, columnspan=3, padx=5, pady=5)

buLoad = tkinter.Button(
    window, text="Load from file", command=load, width=15)
buLoad.grid(row=2, column=0, sticky="w", padx=5, pady=5)

buSave = tkinter.Button(
    window, text="Save to file", command=save, width=15)
buSave.grid(row=2, column=1, padx=5, pady=5)

buDelete = tkinter.Button(
    window, text="Delete text", command=delete, width=15)
buDelete.grid(row=2, column=2, sticky="e", padx=5, pady=5)

buEnd = tkinter.Button(window, text="End", command=end, width=10)
buEnd.grid(row=3, column=3, padx=5, pady=5)

window.mainloop()
```

Listing 10.6 gui_multiline.py File

Figure 10.8 shows the display after calling the program and clicking the **Load from file** button.

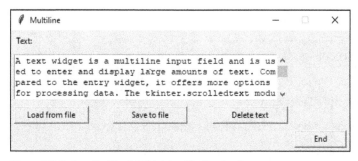

Figure 10.8 ScrolledText Widget with Content

Clicking the **Delete text** button deletes the character string in the ScrolledText widget using the delete() method. In this case, the string is deleted in its entirety using the 1.0 and "end" parameters.

Thanks to the columnspan parameter (which has the value 3 in our example), the ScrolledText widget extends over three columns of the grid, which is created using the *pack* geometry manager. The **Load from file** and **Delete text** buttons are aligned to the left or right edge of their respective cells using the sticky parameter.

When you click the **Load from file** button, the load() function is called. First, the string in the ScrolledText widget is deleted.

Then, an attempt is made to open the text file, with the following possible results:

- If successful, the entire content of the text file will be read using the read() method and inserted into the ScrolledText widget using the insert() method. The first parameter specifies the insertion point. The value 1.0 is used to select the start of the ScrolledText widget. Please note that the text file to be read from must be in American National Standards Institute (ANSI) encoding.
- If opening is not successful, an error message will be displayed in the ScrolledText widget.

Once the **Save to file** button is clicked, the save() function is called. The text file is opened for writing. The content of the ScrolledText widget is determined using the get() method and the 1.0 and "end" parameters (from the first to the last character) and written to the text file using the write() method.

This approach allows you to save a text file on Windows using the default editor in ANSI encoding. Select the **File • Save as** menu item, expand the **Coding** list, select the **ANSI** option, and click the **Save** button, as shown in Figure 10.9.

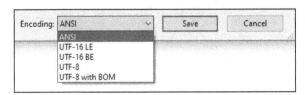

Figure 10.9 Saving in ANSI Coding

10.2.4   List Box with Simple Selection

A listbox widget serves as a selection menu for a series of terms from which one or more can be selected.

The following program creates a list box with a total of ten entries. This list box is linked to a scrollbar that makes it easier to select an entry. The list box and scrollbar are combined in a frame widget that allows these elements to be arranged together.

```python
import tkinter

def output():
    lbOutput["text"] = liSelection.get("active")

def end():
    window.destroy()

window = tkinter.Tk()
window.title("Simple")
window.resizable(0, 0)

lbSelection = tkinter.Label(window, text="Your selection:")
lbSelection.grid(row=0, column=0, sticky="w", padx=5, pady=5)

frSelection = tkinter.Frame(window)
frSelection.grid(row=1, column=0, padx=5, pady=5)

sbSelection = tkinter.Scrollbar(frSelection, orient="vertical")
liSelection = tkinter.Listbox(frSelection, height=4,
    yscrollcommand=sbSelection.set)
sbSelection["command"] = liSelection.yview

city = ["Houston", "Seattle", "Boston", "Detroit", "Tampa",
        "Denver", "Portland", "Hartford", "Fremont", "Atlanta"]
for c in city:
    liSelection.insert("end", c)
liSelection.grid(row=0, column=0)
sbSelection.grid(row=0, column=1, sticky="sn")

buOutput = tkinter.Button(window, text="Output", command=output, width=10)
buOutput.grid(row=2, column=0, sticky="w", padx=5, pady=5)

lbOutput = tkinter.Label(window, text="(empty)")
lbOutput.grid(row=3, column=0, sticky="w", padx=5, pady=5)

buEnd = tkinter.Button(window, text="End", command=end, width=10)
buEnd.grid(row=4, column=2, padx=5, pady=5)

window.mainloop()
```

Listing 10.7 gui_list.py File

After selecting an entry and clicking the **Output** button, the display shown in Figure 10.10 appears.

Figure 10.10 List with Simple Selection Option

A frame widget is created and assigned to the frSelection variable. This widget is not visible and is only used to arrange other widgets.

The scrollbar widget is assigned to the frame widget and to the sbSelection variable. The vertical value for the orient parameter leads to a vertical scrollbar. A scrollbar can be assigned to different widget types.

The listbox widget is assigned to the frame widget and to the liSelection variable. The list box has a height of 4.

Two measures are necessary to establish the connection between the list box and the scrollbar:

1. The yscrollcommand parameter of the list box receives the sbSelection.set value.

2. The command property of the scrollbar is assigned the liSelection.yview value.

Now, the following rule applies: If the scrollbar is clicked, the list box scrolls.

The insert() method adds an element to a listbox widget. The end value selects an insertion point at the end of the list. The list box is filled with the ten elements of the city list.

The grid() method is called for the listbox widget and the scrollbar widget. The specifications for the row and column parameters refer to the surrounding element (i.e., the frame widget). The sn value for the sticky parameter ensures that the scrollbar widget extends to the top and bottom edges of the frame widget (i.e., over the entire height of the list box, which in turn determines the height of the frame widget).

The frame widget determines the width of the column on the left. The two labels and the button are arranged on the left edges of their cells using the sticky parameter.

In the output() function, the get() method is called. This method returns the list element for the specified number, starting at 0. If active is specified, the currently selected element will be provided.

> **Note**
>
> You can add entries to a list box at any position, starting at 0, using a number for the insertion point. For example, if you were to specify the value 0 for all calls of the `insert()` method, each element of the list would be inserted at the start of the list box. The entries would therefore appear in reverse order.

10.2.5 List Box with Multiple Selection Option

You can also select multiple elements in a `listbox` widget. The following program excerpt only shows the differences to our previous program.

```python
...
def output():
    output = ""
    for index in liSelection.curselection():
        output += f"{liSelection.get(index)}\n"
    lbOutput["text"] = output
...
liSelection = tkinter.Listbox(frSelection,
    height=4, yscrollcommand=sbSelection.set, selectmode="multiple")
...
lbOutput = tkinter.Label(window, text="(empty)", height=10, anchor="n")
lbOutput.grid(row=3, column=0, sticky="w", padx=5, pady=5)
...
```

Listing 10.8 gui_list_multiple.py File

After selecting multiple entries and clicking on the **Output** button, the display shown in Figure 10.11 appears.

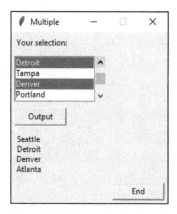

Figure 10.11 List with Multiple Selection Option

The entries are clicked on one after the other for selection. To deselect a selected entry, you need to click on it again.

The list box is also provided with the selectmode property and the multiple property value. The default value is single and therefore did not need to be specified in the previous example.

The curselection() method returns the indexes of the currently selected entries in a tuple. The corresponding entries from the list box are output.

The label widget has a height that would be sufficient to display all elements. The anchor parameter is used to arrange the text at the top of the label widget.

10.2.6   Spin Boxes

Since Python 3.7, the tkinter.ttk.Spinbox class is available for creating a spinbox widget. It corresponds to a combo box.

In a spin box, you can perform the following actions:

- Enter a value as in an entry widget
- Select a value from a range of numbers
- Select an entry from a list

Let's now look at a program in which these options are implemented.

```python
import tkinter.ttk

def selectionCity():
    lbOutput["text"] = "Selection: " + spCity.get()

def selectionNumber():
    lbOutput["text"] = "Selection: " + spNumber.get()

def output():
    lbOutput["text"] = f"Output: {spCity.get()}/{spNumber.get()}"

def end():
    window.destroy()

window = tkinter.Tk()
window.title("Spinbox")
window.resizable(0, 0)

lbSelection = tkinter.Label(window, text="Your input or selection:")
lbSelection.grid(row=0, column=0, padx=5, pady=5)
```

```
city = ["Houston", "Seattle", "Boston", "Detroit", "Tampa",
        "Denver", "Portland", "Hartford", "Fremont", "Atlanta"]
spCity = tkinter.ttk.Spinbox(window, values=city, command=selectionCity)
spCity.set("Houston")
spCity.grid(row=1, column=0, sticky="w", padx=5, pady=5)

spNumber = tkinter.ttk.Spinbox(window,
    from_=10, to=20, width=15, command=selectionNumber)
spNumber.set(12)
spNumber.grid(row=2, column=0, sticky="w", padx=5, pady=5)

buOutput = tkinter.Button(window, text="Output", command=output, width=10)
buOutput.grid(row=3, column=0, sticky="w", padx=5, pady=5)

lbOutput = tkinter.Label(window, text="(empty)")
lbOutput.grid(row=4, column=0, sticky="w", padx=5, pady=5)

buEnd = tkinter.Button(window, text="End", command=end, width=10)
buEnd.grid(row=5, column=1, sticky="w", padx=5, pady=5)

window.mainloop()
```

Listing 10.9 gui_spinbox.py File

Once you have selected an entry in the upper spin box, the display should resemble the
window shown in Figure 10.12.

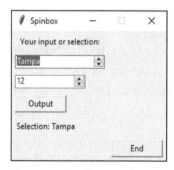

Figure 10.12 Selection of an Entry in the Upper Spin Box

Select an entry in the lower spin box, as shown in Figure 10.13.

After entering a character string in the upper spin box and a number in the lower spin
box and then clicking the **Output** button, the window should resemble the screen
shown in Figure 10.14.

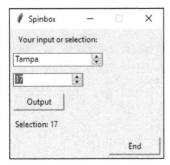

Figure 10.13 Selection of an Entry in the Lower Spin Box

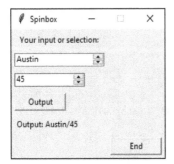

Figure 10.14 Entering New Entries

The elements in the `city` list are displayed in the `spCity` spin box at the top. They were added to the spin box using the `values` parameter. The numbers from 10 to 20 are displayed in the lower spin box. They were added to the spin box using the `from_` and `to` parameters. The `set()` method is used to set the start value of a spin box.

A spin box can be connected to a callback function. This function is called as soon as a selection is made in the spin box. The selection is determined using the `get()` method. However, the callback function is not called after an input in the spin box.

After clicking the **Output** button, the current values of the two spin boxes are displayed. This action applies to both selected and entered values.

10.2.7   Radio Buttons and the Widget Variable

Radio buttons allow for a simple selection, similar to a list box with a simple selection. Radio buttons are individual widgets that are combined into a group so that they respond together. For this purpose, they must be linked to a *widget variable*.

Note

You can link multiple widget types with widget variables. A change in the widget's status leads to a change in the value of the widget variable, and vice versa.

In the following example, the user can select a color using three radio buttons.

```python
import tkinter

def selection():
    lbOutput["text"] = "Selection: " + color.get()

def output():
    lbOutput["text"] = "Output: " + color.get()

def end():
    window.destroy()

window = tkinter.Tk()
window.title("Radiobuttons")
window.resizable(0, 0)

lbSelection = tkinter.Label(window, text="Color:", anchor="w", width=20)
lbSelection.grid(row=0, column=0, sticky="w", padx=5, pady=5)

color = tkinter.StringVar()
color.set("Red")

frColor = tkinter.Frame(window)
frColor.grid(row=1, column=0, padx=5, pady=5)

rbRed = tkinter.Radiobutton(frColor, text="Red", variable=color,
    value="Red", command=selection)
rbRed.grid(row=0, column=0, padx=5, pady=5)

rbYellow = tkinter.Radiobutton(frColor, text="Yellow", variable=color,
    value="Yellow", command=selection)
rbYellow.grid(row=0, column=1, padx=5, pady=5)

rbBlue = tkinter.Radiobutton(frColor, text="Blue", variable=color,
    value="Blue", command=selection)
rbBlue.grid(row=0, column=2, padx=5, pady=5)

buOutput = tkinter.Button(window, text="Output", command=output, width=10)
buOutput.grid(row=2, column=0, sticky="w", padx=5, pady=5)

lbOutput = tkinter.Label(window, text="(empty)")
lbOutput.grid(row=3, column=0, sticky="w", padx=5, pady=5)
```

10

```
buEnd = tkinter.Button(window, text="End", command=end, width=10)
buEnd.grid(row=4, column=1, padx=5, pady=5)

window.mainloop()
```

Listing 10.10 gui_radio.py File

Select the **Yellow** option, as shown in Figure 10.15.

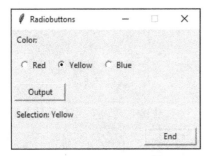

Figure 10.15 Group of Radio Buttons

A widget variable must be an object from one of the following classes:

- StringVar (for character strings)
- IntVar (for integers)
- DoubleVar (for numbers with decimal places)

In our example, the widget variable color is created as an object of the StringVar class. The set() method is used to assign it the start value Red.

The three radio buttons are embedded next to each other in a frame widget. They have the color value for the variable property and are thus linked to each other and to the widget variable (color). If a radio button is selected, the corresponding value of the radio button's value property is assigned to the widget variable.

The widget variable color has the start value Red. Thus, the corresponding radio button is already selected at the start of the program. Such a pre-assignment is part of good programming style and avoids the potential problem of no selection being made at all.

The selection() function is called as soon as one of the radio buttons is selected. The output() function is called after the **Output** button has been clicked. In both functions, the get() method determines the value of the widget variable and therefore the current selection.

Usually, only one of the two functions is required in an application: Either you want the selection of a radio button to immediately trigger another action, or you want to know the status of the group of radio buttons only after all elements of an application have been operated.

10.2.8   Check Buttons

Check buttons allow multiple selections, similar to a list box with multiple selections. You can determine the status of a check button, like the status of a group of radio buttons, either immediately or only after operating all elements.

In our next example, reservations for hotel rooms are made. Using two check buttons, you can decide whether you want a room with a shower (or not) and with a minibar (or not).

```python
import tkinter

def selection():
    lbOutput["text"] = f"Selection: {shower.get()}, {minibar.get()}"

def output():
    lbOutput["text"] = f"Output: {shower.get()}, {minibar.get()}"

def end():
    window.destroy()

window = tkinter.Tk()
window.title("Checkbuttons")
window.resizable(0, 0)

lbSelection = tkinter.Label(window, text="Room:", anchor="w", width=30)
lbSelection.grid(row=0, column=0, padx=5, pady=5)

shower = tkinter.StringVar()
shower.set("without shower")

minibar = tkinter.StringVar()
minibar.set("without minibar")

ckShower = tkinter.Checkbutton(window, text="Shower", variable=shower,
    onvalue="with shower", offvalue="without shower", command=selection)
ckShower.grid(row=1, column=0, sticky="w", padx=5, pady=5)

ckMinibar = tkinter.Checkbutton(window, text="Minibar", variable=minibar,
    onvalue="with minibar", offvalue="without minibar", command=selection)
ckMinibar.grid(row=2, column=0, sticky="w", padx=5, pady=5)

buOutput = tkinter.Button(window, text="Output",
    command=output, width=10)
buOutput.grid(row=3, column=0, sticky="w", padx=5, pady=5)
```

10

```
lbOutput = tkinter.Label(window, text="(empty)")
lbOutput.grid(row=4, column=0, sticky="w", padx=5, pady=5)

buEnd = tkinter.Button(window, text="End", command=end, width=10)
buEnd.grid(row=5, column=2, padx=5, pady=5)

window.mainloop()
```

Listing 10.11 gui_check.py File

Select both options, as shown in Figure 10.16.

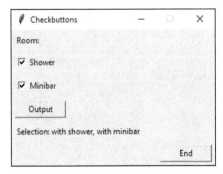

Figure 10.16 After Selecting Both Options

The two widget variables shower and minibar are assigned the initial values "without shower" and "without minibar" using the set() method.

The two check buttons are both linked to the associated variable via the variable property. The onvalue and offvalue properties represent the two states: "Checkbutton is selected" and "Checkbutton is not selected".

The select() function is called as soon as a check button is selected or deselected. The output() function is called after the **Output** button has been clicked. In both functions, the get() method determines the value of the two widget variables and thus the current selection.

10.2.9   Sliders and Scales

A scale widget corresponds to a slider (i.e., the visual display of a value that can be changed using the mouse).

Our next example simulates a speedometer. Speeds between 0 and 200 mph can be displayed and set in increments of 5 mph. The set value is also displayed in a label.

```
import tkinter

def action(self):
```

```
    lbValue["text"] = f"Action: {scValue.get()} mph"

def output():
    lbValue["text"] = f"Output: {scValue.get()} mph"

def end():
    window.destroy()

window = tkinter.Tk()
window.title("Scale")
window.resizable(0, 0)

value = tkinter.IntVar()
value.set(100)

scValue = tkinter.Scale(window, length=300, orient="horizontal",
    from_=0, to=200, resolution=5, tickinterval=20, label="mph",
    command=action, variable=value)
scValue.grid(row=0, column=0, padx=5, pady=5)

buOutput = tkinter.Button(window, text="Output", command=output, width=10)
buOutput.grid(row=1, column=0, sticky="w", padx=5, pady=5)

lbValue = tkinter.Label(window, text="Start: 100 mph")
lbValue.grid(row=2, column=0, sticky="w", padx=5, pady=5)

buEnd = tkinter.Button(window, text="End", command=end, width=10)
buEnd.grid(row=3, column=1, padx=5, pady=5)

window.mainloop()
```

Listing 10.12 gui_scale.py File

After activating the slider, your screen should look like the one shown in Figure 10.17.

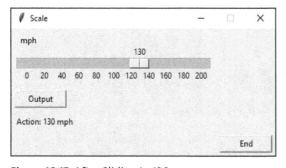

Figure 10.17 After Sliding to 130

The widget variable value is preset with the value 100.

The scale widget is aligned horizontally by using the orientation property, has a label field with the value "mph" thanks to the label property, and is given a length of 300 via the length property. Due to the from_ and to properties, this widget displays values from 0 to 200.

The value for the resolution property indicates the increment when sliding. The tick-interval property defines the distance between the displayed values. The slider is linked to the widget variable value via the variable property.

The action() function is called when the slider is operated, and the output() function is called when the **Output** button is clicked. In both functions, the current value of the slider is determined and displayed using the get() method.

10.3   Images and Mouse Events

In this section, I will explain how you can use images and mouse events in the application window.

10.3.1   Embedding and Changing an Image

The content of an image file can be displayed on a label by using an object of the Photo-Image class. The properties of such an object can be determined and changed.

In this example, the width and height of an image are specified. The color and transparency of individual pixels are also determined and changed.

```
import tkinter

def font():
    for x in range(5, 94):
        for y in range(69, 80):
            imImage.put("#00549D", (x, y))

def transparent():
    for x in range(5, 94):
        for y in range(38, 46):
            imImage.transparency_set(x, y, True)
    if imImage.transparency_get(50, 40):
        print("Point is transparent")

def end():
    window.destroy()
```

```
window = tkinter.Tk()
window.title("Image")
window.resizable(0, 0)

imImage = tkinter.PhotoImage(file="rheinwerk.png")
lbImage = tkinter.Label(window)
lbImage["image"] = imImage
lbImage.grid(row=0, column=0, padx=5, pady=5)

print("Width:", imImage.width())
print("Height:", imImage.height())
print("Color x,y:", imImage.get(0, 0))
if not imImage.transparency_get(50, 40):
    print("Point is not transparent")

buFont = tkinter.Button(window,
    text="Font in blue", command=font, width=20)
buFont.grid(row=1, column=0, padx=5, pady=5)

buTransparent = tkinter.Button(window,
    text="Area transparent", command=transparent, width=20)
buTransparent.grid(row=2, column=0, padx=5, pady=5)

buEnd = tkinter.Button(window, text="End", command=end, width=10)
buEnd.grid(row=3, column=1, padx=5, pady=5)

window.mainloop()
```

Listing 10.13 gui_image.py File

Call the program, and you should see the window shown in Figure 10.18.

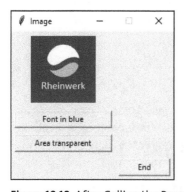

Figure 10.18 After Calling the Program

327

After clicking the **Font in blue** button, the lettering in the logo is deleted by changing the color of the pixels in this area to blue, as shown in Figure 10.19.

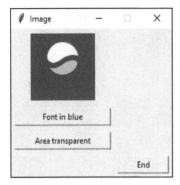

Figure 10.19 After Deleting the Lettering

After clicking the **Area transparent** button, the pixels in an area become transparent, as shown in Figure 10.20.

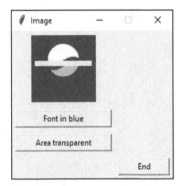

Figure 10.20 After Changing the Transparency

In addition, the following information appears in the **IDLE Shell**:

```
Width: 100
Height: 100
Color x,y: (0, 84, 157)
Point is not transparent
```

The image object is transferred to the application as an object of the PhotoImage class. The file parameter contains the name of the image file whose content is displayed. The display occurs in a label. The reference to the image object is assigned to the image property of the label.

The width() and height() methods return the width and height of the image in pixels, and the get() method returns the color of the specified pixel as a tuple of three values between 0 and 255 for the red, green, and blue components of the RGB color model. The two coordinates (x and y) of the pixel are passed as parameters. The coordinates 0, 0 represent the top-left corner.

The `transparency_get()` method has been available since Python 3.8. This method provides information as to whether the pixel in question is transparent.

After clicking the **Font in blue** button, the `put()` method is called in the `font()` function for all pixels in an area. This function is used to set the color of the pixel. The new color is noted as the first parameter using a Hypertext Markup Language (HTML) color specification in the #RRGGBB format.

The # character is followed by two hexadecimal digits for the red, green, and blue color components. The second parameter is a tuple with the two coordinates for x and y. Hexadecimal digits are (in ascending order): 0, 1, 2, 3, 4, 5, 6, 7, 8, 9, A (corresponds to the decimal value 10), B (= 11), C (= 12), D (= 13), E (= 14), F (= 15); see also Chapter 4, Section 4.1.1.

In this example, a double loop is used to color a strip of pixels in the color #00549D, which is the standard dark blue of the Rheinwerk logo. In its decimal form, the value corresponds to 0, 84, 157. The "Rheinwerk" lettering is then no longer visible.

After clicking the **Area transparent** button, the `transparent()` function calls the `transparency_set()` method, which has also been available since Python 3.8. This method is used to set the transparency of a pixel. The first two parameters indicate the two coordinates for x and y. The third parameter specifies whether the point should be transparent (`True`) or not (`False`). In this example, a strip of pixels in the middle of the image is made transparent. The corresponding information is output for every pixel in this strip.

10.3.2 Mouse Events

The widgets that have led to actions in the previous programs contain the `command` property. This property links the widget to a function that is called after the most frequently used event of the widget in question, usually a click event.

You can also link widgets to other mouse events. Some of these events, with their corresponding names, as required for your program in parentheses, include the following:

- Clicking the left or right mouse button (`<Button1>` or `<Button3>`)
- Releasing the left or right mouse button (`<ButtonRelease 1>` or `<ButtonRelease 3>`)
- Moving the mouse into a widget (`<Enter>`)
- Moving the mouse within a widget (`<Motion>`)
- Moving the mouse out of a widget (`<Leave>`)

These options are implemented in our next program.

```python
import tkinter

def moved(e):
    lbOutput["text"] = f"x={e.x}, y={e.y}"

def left(e):
    lbOutput["text"] = "Left pressed"

def leftrelease(e):
    lbOutput["text"] = "Left released"

def right(e):
    lbOutput["text"] = "Right pressed"

def rightrelease(e):
    lbOutput["text"] = "Right released"

def leave(e):
    lbOutput["text"] = "Leave"

def enter(e):
    lbOutput["text"] = "Enter"

def end():
    window.destroy()

window = tkinter.Tk()
window.title("Mouse")
window.resizable(0, 0)

imImage = tkinter.PhotoImage(file="rheinwerk.png")
lbImage = tkinter.Label(window, image=imImage)
lbImage.bind("<Motion>", moved)
lbImage.bind("<Button 1>", left)
lbImage.bind("<ButtonRelease 1>", leftrelease)
lbImage.bind("<Button 3>", right)
lbImage.bind("<ButtonRelease 3>", rightrelease)
lbImage.bind("<Leave>", leave)
lbImage.grid(row=0, column=0, padx=5, pady=5)

lbOutput = tkinter.Label(window, text="(empty)")
lbOutput.grid(row=1, column=0, sticky="w", padx=5, pady=5)

buEnd = tkinter.Button(window, text="End", command=end, width=10)
```

```
buEnd.grid(row=2, column=1, padx=5, pady=5)
buEnd.bind("<Enter>", enter)

window.mainloop()
```

Listing 10.14 gui_mouse.py File

As shown in Figure 10.21, the mouse is located in the label with the image.

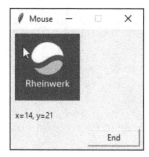

Figure 10.21 Mouse in the Top-Left Area

As you release your right mouse button in the widget, you should see the message shown in Figure 10.22.

Figure 10.22 Right Mouse Button Released

A total of seven functions are defined, which are linked to certain events and widgets. In this case, the "Enter" event is linked to the **End** button, while the other events are linked to the label in which an image is displayed. Information about the respective event is displayed in another label.

The bind() method creates the connection. The first parameter of the method is a character string with the name of the event. The second parameter is the name of the function that is called when the event occurs.

Each event function transmits information about the event using an event object (in our example, e). For a mouse event, the object properties x and y are of interest because they provide the current coordinates of the mouse pointer within the widget.

> **Differences on macOS**
>
> On macOS, the event <Button 3> has no effect, because right-clicking is handled differently.

10.4   The place Geometry Manager

The *place* geometry manager enables the arrangement of widgets via coordinates. In this case, the size of the application window does not depend on the widgets it contains and must therefore be set.

10.4.1   Window Size and Absolute Position

In the next example, we'll create the GUI from Section 10.1.2 using the *place* geometry manager.

```python
import tkinter

def hello():
    lbOutput["text"] = "Hello"

def end():
    window.destroy()

window = tkinter.Tk()
window.title("GUI")
window.geometry("245x105+200+50")
window.resizable(0, 0)

buHello = tkinter.Button(window, text="Hello", command=hello, width=10)
buHello.place(x=5, y=5)

lbOutput = tkinter.Label(window, text="(empty)",
    anchor="w", relief="sunken", width=20)
lbOutput.place(x=5, y=40)

buEnd = tkinter.Button(window, text="End", command=end, width=10)
buEnd.place(x=160, y=75)

window.mainloop()
```

Listing 10.15 gui_place.py File

After clicking the **Hello** button, the application looks as shown in Figure 10.23.

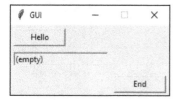

Figure 10.23 Window Size and Absolute Position

You can use the geometry() method to define the size and position of the application window. The first two numbers are connected with the "x" character and indicate the width and height of the window. The next two numbers each follow the "+" sign and indicate the distance between the top-left corner of the window and the top-left corner of the screen. If you only specify the size, for example, via window.geometry("245x105"), then the window is arbitrarily placed.

When the place() method is called, values must be transferred for the named x and y parameters. They determine the distance between the top-left corner of the widget and the top-left corner of the application window area below the title bar.

As the size of the application window does not depend on the widgets it contains, its final size and the position of the widgets are usually only achieved after repeated trial and error.

10.4.2 Relative Position

Using the *place* geometry manager, you can also select the position of a widget in relation to the size of the application window. This capability is particularly advantageous if the size of the window can be changed.

In the next example, a total of nine buttons, labeled 1 through 9, are positioned in the corners or in the middle of the edges of the application window. The buttons are not linked to an action and only serve as examples for any widgets. The window can be enlarged or reduced within defined values.

```
import tkinter

window = tkinter.Tk()
window.title("GUI")
window.geometry("300x225")
window.minsize(200, 150)
window.maxsize(400, 300)

bu1 = tkinter.Button(window, text="1", width=5)
```

```
bu1.place(relx=0.05, rely=0.05)

bu2 = tkinter.Button(window, text="2", width=5)
bu2.place(relx=0.5, rely=0.05, anchor="n")

bu3 = tkinter.Button(window, text="3", width=5)
bu3.place(relx=0.95, rely=0.05, anchor="ne")

bu4 = tkinter.Button(window, text="4", width=5)
bu4.place(relx=0.05, rely=0.5, anchor="w")

bu5 = tkinter.Button(window, text="5", width=5)
bu5.place(relx=0.5, rely=0.5, anchor="center")

bu6 = tkinter.Button(window, text="6", width=5)
bu6.place(relx=0.95, rely=0.5, anchor="e")

bu7 = tkinter.Button(window, text="7", width=5)
bu7.place(relx=0.05, rely=0.95, anchor="sw")

bu8 = tkinter.Button(window, text="8", width=5)
bu8.place(relx=0.5, rely=0.95, anchor="s")

bu9 = tkinter.Button(window, text="9", width=5)
bu9.place(relx=0.95, rely=0.95, anchor="se")

window.mainloop()
```

Listing 10.16 gui_relative.py File

After starting the program, the application displays the screen shown in Figure 10.24.

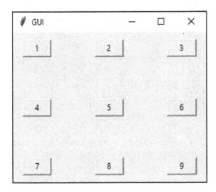

Figure 10.24 Relative Position after the Start

In the smallest possible size, the application is displayed in the screen shown in Figure 10.25.

Figure 10.25 Relative Position after Downsizing

The minsize() and maxsize() methods are used to define the minimum and maximum size of the application window.

Values between 0 (far left or far top) and 1 (far right or far bottom) can be selected for the relx and rely parameters. If the anchor parameter is not specified, the top-left corner of the widget is selected as the reference point for the relative position. Other points of reference include the following options:

- n: Center of the top of the widget
- ne: Top-right corner of the widget
- e: Center of the right edge of the widget
- se: Bottom-right corner of the widget
- s: Center of the bottom of the widget
- bw: Bottom-left corner of the widget
- w: Center of the left edge of the widget
- center: Center of the whole widget

10.4.3   Changing the Position

Calling one of the geometry managers at application runtime allows you to change the position of a widget.

In our next example, a widget can be moved to the left or right using two buttons and the place() method.

```
import tkinter, time

posx = 205

def left():
    global posx
    if posx > 80:
```

```
        posx -= 20
        buMove.place(x=posx, y=5)

def right():
    global posx
    if posx < 340:
        posx += 20
        buMove.place(x=posx, y=5)

window = tkinter.Tk()
window.title("Move")
window.geometry("455x35")
window.resizable(0, 0)

buLeft = tkinter.Button(window, text="<<", command=left, width=5)
buLeft.place(x=5, y=5)

buMove = tkinter.Button(window, text="xxx", width=5)
buMove.place(x=205, y=5)

buRight = tkinter.Button(window, text=">>", command=right, width=5)
buRight.place(x=405, y=5)

window.mainloop()
```

Listing 10.17 gui_move.py File

After clicking the **>>** button multiple times, you'll see the window shown in Figure 10.26.

Figure 10.26 Changing the Position

The posx variable is used to save the current position of the widget with the **xxx** label in the x direction. The start position 205 is saved at the beginning.

Clicking the two buttons leads to the left() and right() functions. In this case, the posx variable is first adopted as global. The system checks whether the widget is already on the far left or far right. If not, the value of the posx variable will be changed. The widget is then set to the corresponding position, which is a little further to the left or right of the current position, using the place() method.

10.5   Menus and Dialog Boxes

In this section, I present menus, standard dialog boxes, and custom dialog boxes. In addition to the widgets within the application window, many GUIs also have control panels of the following types:

- A menu bar that is permanently visible at the top edge
- Context menus that are assigned to individual widgets and are only displayed when required

Python provides Menu widgets for this purpose.

10.5.1   Menu Bars

The creation of a menu bar with individual menus and the menu commands they contain requires the following steps:

1. Creating an object of the Menu class as a menu bar
2. Creating additional objects of the Menu class as individual menus within the menu bar
3. Creating menu items within a menu
4. Adding the menus to the menu bar and labeling the menus
5. Adding the menu bar to the window

The next program contains a menu bar with two menus, each of which contains several menu items of different types.

```python
import tkinter

def colorChange():
    lbHello["bg"] = color.get()

def edgeChange():
    lbHello["relief"] = edge.get()

def end():
    window.destroy()

window = tkinter.Tk()
window.title("Menu")
window.resizable(0, 0)

lbHello = tkinter.Label(window, text="Hello", width=30, bg="#FAAC58")
lbHello.grid(row=0, column=0, padx=20, pady=40)
```

```
mBar = tkinter.Menu(window)

mFile = tkinter.Menu(mBar)
mFile.add_command(label="New")
mFile.add_command(label="Open")
mFile.add_separator()
mFile.add_command(label="End", command=end)

mView = tkinter.Menu(mBar)
mView["tearoff"] = 0

color = tkinter.StringVar()
color.set("#FAAC58")
mView.add_radiobutton(label="Brown", variable=color,
    value="#FAAC58", underline=0, command=colorchange)
mView.add_radiobutton(label="Green", variable=color,
    value="#ACFA58", underline=0, command=colorchange)
mView.add_separator()

edge = tkinter.StringVar()
edge.set("flat")
mView.add_checkbutton(label="Edge", variable=edge,
    onvalue="solid", offvalue="flat", underline=0, command=edgeChange)

mBar.add_cascade(label="File", menu=mFile)
mBar.add_cascade(label="View", menu=mView)
window["menu"] = mBar

window.mainloop()
```

Listing 10.18 gui_menu.py File

The **File** menu is shown in Figure 10.27. Only the **End** menu item is assigned a function, while the other menu items are for display purposes only.

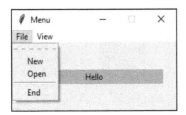

Figure 10.27 The File Menu

The **View** menu is shown in Figure 10.28. Clicking on one of the two menu items changes the background color or the edge of the label.

A dashed line is displayed above each menu by default, as in the **File** menu. Clicking on this line causes a *tear-off* of the menu in question, which means it will be swapped out as a separate window. This option is suppressed in the **View** menu.

The label is assigned the background color #FAAC58 for light brown via the bg property (short for *background*).

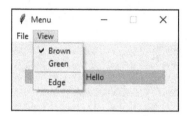

Figure 10.28 The View Menu

An object of the Menu class is then created as a menu bar for the window. This object can be accessed via the mBar variable. The **File** menu is created within the menu bar, which can be accessed via the mFile variable.

The add_command() method creates a standard menu item, which leads to the execution of a command. The label property is used for the visible display of the menu item. The add_separator() method creates a visual separator between thematically separated menu items.

Within the menu bar, the second menu named **View** is created, which can be accessed via the mView variable. The tearoff property is set to 0 so that the menu cannot be disconnected.

The widget variable color is created for selecting the background color of the label, with the appropriate start value of #FAAC58.

The add_radiobutton() method creates two menu items as a related group, which react similarly to radio buttons in the application window. The common connection to the widget variable color is created via the variable parameter. The value of the value property stands for the desired color. After clicking one of the menu items, the colorChange() function is called, in which the currently selected color value is determined using the get() method and used to set the background color.

The underline property defines the index of the underlined letter of a menu item. After selecting the menu by pressing ⎇Alt⎇, simply enter the corresponding letter to select the menu item.

The widget variable edge is created for setting the edge of the label, with the appropriate start value, flat.

The add_checkbutton() method is used to create a menu item that reacts similarly to a check button in the application window. The connection to the widget variable edge is created via the variable parameter. The values of the onvalue and offvalue properties stand for the setting of the edge. After clicking the menu item, the edgeChange() function is called, in which the get() method is used to determine whether the menu item is selected or not. The edge then is set accordingly.

The two previously created menus—**File** and **View**—are added to the menu bar using the add_cascade() method. The menu property contains a reference to the respective menu object. The menu bar is added to the application window via the menu property.

> **Difference on macOS**
>
> The menus appear in the Mac menu bar, as is usual on a Mac.

10.5.2   Context Menus

You can create a context menu in a similar way to a menu bar by following these steps:

1. Creating an object of the Menu class as a context menu
2. Creating menu items within the context menu
3. Connecting an event with a widget, usually with a right-click
4. As a result of this event, a context menu is displayed near the relevant widget

In the next program, the background color of a label can be changed via its context menu.

```
import tkinter

def showContext(e):
    mContext.tk_popup(e.x_root, e.y_root)

def colorChange():
    lbHello["bg"] = color.get()

def end():
    window.destroy()

window = tkinter.Tk()
window.title("Context menu")
window.resizable(0, 0)

lbHello = tkinter.Label(window, text="Hello", width=20, bg="#FAAC58")
```

```
lbHello.bind("<Button 3>", showContext)
lbHello.grid(row=0, column=0, padx=20, pady=40)

color = tkinter.StringVar()
color.set("#FAAC58")

mContext = tkinter.Menu(window)
mContext["tearoff"] = 0
mContext.add_radiobutton(label="Brown", variable=color,
    value="#FAAC58", command=colorchange)
mContext.add_radiobutton(label="Green", variable=color,
    value="#ACFA58", command=colorchange)

buEnd = tkinter.Button(window, text="End", command=end, width=10)
buEnd.grid(row=0, column=1, padx=20, pady=20)

window.mainloop()
```

Listing 10.19 gui_context.py File

After right-clicking within the label, you'll see the context menu shown in Figure 10.29.

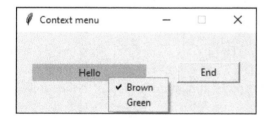

Figure 10.29 Context Menu for the Label

The context menu is created like a standard menu in the usual way but is not displayed.

By right-clicking in the label, the showContext() function is called. An event object is transferred that contains the coordinates of the mouse click in relation to the application window in the x_root and y_root properties, among others.

In the showContext() function, the tk_popup() method is called for the context menu, which leads to the context menu being displayed. The coordinates of the mouse click are transferred. This position then applies to the top-left corner of the context menu displayed.

If one of the two menu items in the context menu is called, the color in the label is set using the colorChange() function.

> **Differences on macOS**
>
> Since right-clicking is handled differently, the program must be changed. The context menu should appear when the left mouse button is clicked. The relevant binding statement must therefore be changed accordingly with `lbHello.bind("<button 1>", showContext)`.

10.5.3   Standard Dialog Boxes

A message box is a ready-made standard dialog box for sending messages. You can call it by using one of the functions of the `tkinter.messagebox` module. The different types of message boxes have the following tasks:

- `showinfo()` is for transmitting information.
- `showwarning()` is for issuing a warning.
- `showerror()` is for reporting an error.
- `askyesno()` is for asking for a "Yes" or "No" answer.
- `askokcancel()` is for asking for an "OK" or "Cancel" response (not shown in our sample program).
- `askretrycancel()` is for asking for a "Retry" or "Cancel" response (not shown in our sample program).
- `show()` is for requesting an answer with an individual selection of buttons.

In our sample program, each function is called via a separate button. If the operation of the message box results in an answer, a corresponding output appears in the label of the application window.

```python
import tkinter, tkinter.messagebox as mb

def info():
    mb.showinfo("Box","Info")

def warning():
    mb.showwarning("Box", "Warning")

def error():
    mb.showerror("Box", "Error")

def yesno():
    answer = mb.askyesno("Box", "Yes or No")
    lbOutput["text"] = "Yes" if answer == 1 else "No"
```

```python
def question():
    msg_box = mb.Message(window, type=mb.ABORTRETRYIGNORE, icon=mb.QUESTION,
        title="Box", message="Abort, Retry or Ignore")
    answer = msg_box.show()
    if answer == "abort":
        lbOutput["text"] = "Abort"
    elif answer == "retry":
        lbOutput["text"] = "Retry"
    else:
        lbOutput["text"] = "Ignore"

def end():
    window.destroy()


window = tkinter.Tk()


buInfo = tkinter.Button(window, text="Info", width=20, command=info)
buInfo.grid(row=0, column=0, padx=5, pady=5)


buWarning = tkinter.Button(window, text="Warning", width=20, command=warning)
buWarning.grid(row=0, column=1, padx=5, pady=5)


buError = tkinter.Button(window, text="Error", width=20, command=error)
buError.grid(row=0, column=2, padx=5, pady=5)


buYesNo = tkinter.Button(window, text="Yes/No", width=20, command=yesno)
buYesNo.grid(row=1, column=0, padx=5, pady=5)


buQuestion = tkinter.Button(window, text="General question",
    width=20, command=question)
buQuestion.grid(row=1, column=1, padx=5, pady=5)


lbOutput = tkinter.Label(window, text="(empty)")
lbOutput.grid(row=1, column=2, padx=5, pady=5)


buEnd = tkinter.Button(window, text="End", width=20, command=end)
buEnd.grid(row=2, column=2, padx=5, pady=5)


window.mainloop()
```

Listing 10.20 gui_message.py File

How the various message boxes are displayed in Windows is shown in Figure 10.30 through Figure 10.34.

Figure 10.30 Information, with Confirmation

Figure 10.31 Warning, with Confirmation

Figure 10.32 Error, with Confirmation

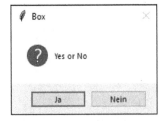

Figure 10.33 Answering "Yes" or "No"

Figure 10.34 Answering a General Question

The called functions with names beginning in show or ask each have two parameters. The first parameter contains the text for the title of the box. The second parameter contains the information text or the question next to the buttons. In addition, suitable icons are displayed and, if necessary, the relevant system sound is output.

Using the functions with names beginning in ask, you create a question and receive a return value. After clicking the left button, the value is 1; after clicking the right button, the value is 0.

The general show() function is not as convenient as the other functions but provides more design options. First, an object of the Message class is created whose properties are assigned several values: type (which buttons), icon (the icon displayed), title (the title), and message (the informational text next to the buttons). Then, the show() function is called for this object and provides a string that is used to identify the button that has been clicked.

10.5.4 Custom Dialog Boxes

You can also create additional dialog boxes and make your GUI application more versatile and clearer. However, you should make sure that only the desired windows and elements can be operated.

In the next program, a second dialog box can be opened. This action deactivates the widgets in the main window; those widgets are reactivated once the second dialog box has been closed.

```
import tkinter

def createTwo():
    buTwo["state"] = "disabled"
    buEnd["state"] = "disabled"
    global windowTwo
    windowTwo = tkinter.Toplevel(window)
    windowTwo.title("Two")
    lbHello = tkinter.Label(windowTwo, text="Hello", width=10)
    lbHello.grid(row=0, column=0, padx=20, pady=10)
    buEndTwo = tkinter.Button(windowTwo, text="End Two",
        width=10, command=endTwo)
    buEndTwo.grid(row=0, column=1, padx=20, pady=10)

def endTwo():
    buTwo["state"] = "normal"
    buEnd["state"] = "normal"
    windowTwo.destroy()
```

```
def end():
    window.destroy()

window = tkinter.Tk()
window.title("Window")
window.resizable(0, 0)

buTwo = tkinter.Button(window, text="Two", width=10, command=createTwo)
buTwo.grid(row=0, column=0, padx=20, pady=10)

buEnd = tkinter.Button(window, text="End", width=10, command=end)
buEnd.grid(row=0, column=1, padx=20, pady=10)

window.mainloop()
```

Listing 10.21 gui_window.py File

Figure 10.35 shows the application window and the second dialog box.

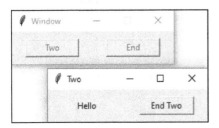

Figure 10.35 Application Window and Second Dialog Box

After clicking the **Two** button, the createTwo() function is called. The two widgets in the main window are deactivated. The second dialog box is created using the Toplevel() function. A reference to it is saved in the global windowTwo variable. A label and a button are created in the second dialog box.

After clicking the **End Two** button, the endTwo() function is called, which reactivates the two widgets in the main window and closes the second dialog box.

10.6   Drawings and Animations

Drawings can be created using a canvas object. The individual objects in a drawing can be displayed, moved, and animated on this canvas.

10.6.1   Various Drawing Objects

In this next program, various objects are depicted in a drawing, as shown in Figure 10.36.

```
import tkinter

window = tkinter.Tk()
window.title("Canvas, Objects")
window.geometry("400x200")
window.resizable(0, 0)

cv = tkinter.Canvas(window, bg="#E0E0E0")
cv.pack(fill="both", expand=True)
cv.create_line((20, 50), (70, 50), (70, 150),
    fill="#A0A0A0", width=3, arrow="last")
cv.create_rectangle((90, 50), (140, 150),
    fill="#A0A0A0", outline="#FFFFFF", width=3)
cv.create_oval((160, 50), (210, 150), fill="#A0A0A0", width=0)
cv.create_polygon((230, 50), (280, 50), (280, 150), fill="#A0A0A0")
cv.create_text(300, 50, text="Hello", anchor="nw")

window.mainloop()
```

Listing 10.22 canvas_objects.py File

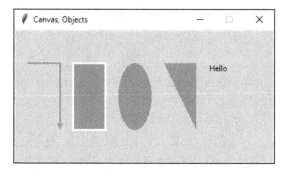

Figure 10.36 Drawing Objects

A reference to the newly created canvas object is transferred to the cv variable. The object has a light gray drawing area as a background.

The canvas object should extend over the entire GUI. The following options are available for the fill parameter:

- The default value none makes sure that a widget cannot be expanded.
- The x value causes the widget to fill the surrounding element horizontally, the y value vertically, while the both value, as in our example, fills it in both dimensions.

347

The True value for the expand parameter enables a widget to occupy all the space that is not taken by other widgets. The default value is False.

The create_line() method is used to create a line. This line consists of straight connections between any number of points. Each point is specified using a pair of coordinates: an x-value and a y-value. The x-value is measured from the left edge of the canvas object, the y-value from the top edge.

Since Python 3.12, the coordinate pairs can be summarized in parentheses for better readability.

The following optional parameters are available:

- The fill and width parameters are used to set the color and thickness of the line. The default values are Black and 1.

- The arrow parameter is required if a line is supposed to have an arrowhead at the beginning or end. You can use the first, last, and both values to indicate the start of the line, the end of the line, or both. If this parameter is omitted, the line has no arrowhead.

The create_rectangle() method is used to draw a rectangle. If the sides of the rectangle have the same length, it is a square. To create the rectangle, the coordinates of the top-left corner and the bottom-right corner are specified.

The following optional parameters are available:

- The fill parameter is used to set the fill color of the rectangle. It is not filled in with a color by default.

- The outline and width parameters set the color and thickness of the border. As with a line, the default values are Black and 1. By using width=0, you can ensure that the rectangle has no border.

The create_oval() method creates an oval within the coordinates of a surrounding rectangle. If the sides of the surrounding rectangle have the same length, the shape is a circle. The optional parameters correspond to those of a rectangle.

The create_polygon() method is used to draw a polygon such as a triangle, for example. The edges of the polygon are specified like a line using any number of points from x and y coordinates. The optional parameters correspond to those of a rectangle. However, a polygon is filled in with a color by default.

The create_text() method creates a text field at a point that is specified using the coordinates for x and y. The text parameter contains the displayed text. The reference point for the coordinates is the center of the text field by default. You can use the optional anchor parameter to set a different reference point. In this context, the nw value makes the top-left corner of the text field the reference point.

The methods each return a reference to the newly created object. These references are not required in our scenario and are therefore not saved.

10.6.2 Controlling Drawing Objects

The move() method is used to move a drawing object, while the coords() method returns the current position of a drawing object. The bind_all() method binds a keyboard event to a widget.

In the following program, a rectangle can be moved on a canvas using the WASD keys, that is, by pressing �W, Ⓐ, Ⓢ, and Ⓓ. The current position of the rectangle is displayed in a text field at the beginning and after each keystroke, as shown in Figure 10.37.

```python
import tkinter

def position():
    x0, y0, x1, y1 = cv.coords(rectangle)
    tx = f"{int(x0)} {int(y0)} {int(x1)} {int(y1)}"
    cv.itemconfigure(output, text=tx)

def key(e):
    if e.char == "w":
        cv.move(rectangle, 0, -10)
    elif e.char == "a":
        cv.move(rectangle, -10, 0)
    elif e.char == "s":
        cv.move(rectangle, 0, 10)
    elif e.char == "d":
        cv.move(rectangle, 10, 0)
    position()

window = tkinter.Tk()
window.title("Canvas, Keys")
window.geometry("400x200")
window.resizable(0, 0)

cv = tkinter.Canvas(window, bg="#E0E0E0")
cv.bind_all("<Key>", key)
cv.pack(fill="both", expand=True)
rectangle = cv.create_rectangle((180, 80), (230, 130),
    fill="#A0A0A0", outline="#A0A0A0")
output = cv.create_text((20, 20), text="", anchor="nw")
position()

window.mainloop()
```

Listing 10.23 canvas_keys.py File

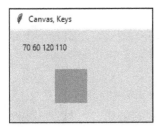

Figure 10.37 Moving Drawing Objects, Outputting the Position

The "Key is pressed in canvas object" event is connected to the callback function key() using the bind_all() method. A rectangle is then created on the canvas. A reference to the rectangle is saved in the rectangle variable. This step is followed by the creation of a text field on the canvas. A reference to this is saved in the output variable. Finally, the starting position of the rectangle is output using the position() method.

An event object containing information about the event is passed to the key() function. The char property of the event object contains the character that was pressed on the keyboard.

Depending on this character, the move() method for the canvas object is called in different ways. The reference of the drawing object to be moved is noted as the first parameter. Two values follow for the displacement in the x-direction and in the y-direction. Note that the y-value must be reduced for an upward shift.

Finally, the new position of the rectangle is output using the position() function.

In the position() function, the coords() method is called for the canvas object. This method expects a reference to the drawing object whose position is supposed to be determined as a parameter. The current position of the drawing object is returned. In the case of a rectangle, this return value is a tuple of four float values. The four values correspond to the values that are required when you create a rectangle. They are converted into integers and combined into a text.

By calling the itemconfigure() method for the canvas object, the properties of a drawing object can be changed. In this case, the text property of the text field is given new content.

10.6.3   Animating Drawing Objects

The after() method of the application window is required for the delayed call of a function.

Such a call is made several times in our next program. This approach enables the animated movement of a drawing object from a starting point to an end point. In addition, the current position of the rectangle is displayed in a text field, as shown in Figure 10.38.

```
import tkinter

def animate():
    x0, y0, x1, y1 = cv.coords(rectangle)
    tx = f"{int(x0)}"
    cv.itemconfigure(output, text=tx)
    if x0 < 330:
        cv.move(rectangle, 2, 0)
        window.after(10, animate)

window = tkinter.Tk()
window.title("Canvas, Animation")
window.geometry("400x150")
window.resizable(0, 0)

cv = tkinter.Canvas(window, bg="#E0E0E0")
cv.pack(fill="both", expand=True)

rectangle = cv.create_rectangle((20, 50), (70, 100),
    fill="#A0A0A0", outline="#A0A0A0")
output = cv.create_text((20, 20), text="", anchor="nw")

window.after(100, animate)
window.mainloop()
```

Listing 10.24 canvas_animation.py File

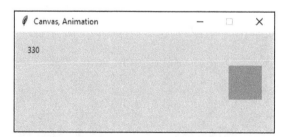

Figure 10.38 Animated Displacement

After creating a rectangle and a text field, the after() method is called for the first time. The first parameter is the delay in milliseconds; the second parameter is the name of the function (animate) whose call is delayed.

The position of the rectangle is determined in the animate() function using the coords() method. The returned tuple is unpacked into four variables. Only a shift to the right takes place in the program, so only the x-values of the position change. The x-value of the top-left corner is output.

If the end point of the shift has not yet been reached, the rectangle is shifted a little to the right. The animate() function is then called again after 10 milliseconds using the after() method.

10.6.4   Collision of Drawing Objects

Multiple drawing objects can be partially or completely superimposed. Drawing objects can move in relation to each other with the help of animations. In games, an important task for a program is to recognize the collision of two drawn objects (i.e., the start of overlapping).

We have two rectangles in the next program. The first rectangle moves towards the second rectangle. The moment the areas of the two rectangles overlap, a collision occurs, and another action takes place.

The find_overlapping() method of the canvas object can be used to detect such a collision, as shown in Figure 10.39.

```python
import tkinter

def animate():
    x0, y0, x1, y1 = cv.coords(rectangle)
    match = cv.find_overlapping(x0, y0, x1, y1)
    if len(match) < 2:
        cv.move(rectangle, 2, 0)
        window.after(10, animate)

window = tkinter.Tk()
window.title("Canvas, Collision")
window.geometry("400x150")
window.resizable(0, 0)

cv = tkinter.Canvas(window, bg="#E0E0E0")
cv.pack(fill="both", expand=True)

rectangle = cv.create_rectangle((20, 50), (70, 100),
    fill="#A0A0A0", outline="#A0A0A0")
cv.create_rectangle((220, 50), (270, 100),
    fill="#606060", outline="#606060")

window.after(100, animate)
window.mainloop()
```

Listing 10.25 canvas_collision.py File

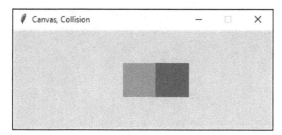

Figure 10.39 Collision of Two Drawing Objects

In the animate() function, the position of the animated rectangle is determined using the coords() method. The returned tuple is unpacked into four variables.

The find_overlapping() method creates a search rectangle using four parameters. This method returns a tuple that contains references of all drawing objects that are located within the search rectangle. If the four parameters are the coordinates of the animated rectangle, the returned tuple contains the references of all drawing objects that collide with the animated rectangle.

One element of the tuple is the reference to the search rectangle itself. If the number of references is less than 2, no collision has taken place, and the animated rectangle can be moved further. If the number of references is 2 or higher, at least one collision has taken place. As a result of this event, the animated rectangle is no longer moved.

10.7 Our Game: GUI Version

Now, let's implement our mental calculation game with a GUI. After starting the program, the application window appears, as shown in Figure 10.40. Enter your name. Then, click the **Start** button.

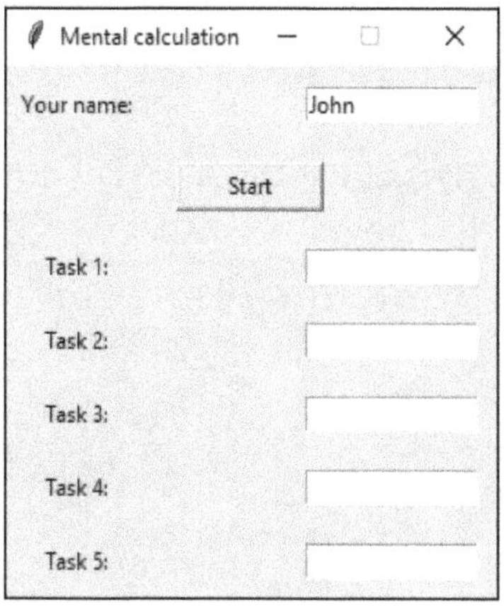

Figure 10.40 After the Start

Then five tasks appear, as shown in Figure 10.41. The button label changes to **Stop**. Determine the five solutions in your head and enter them in the input fields. Then, click the **Stop** button.

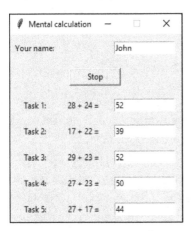

Figure 10.41 Tasks and Results

Now follows the evaluation of your inputs. Once you have solved all the tasks correctly, a highscore list will be displayed, as shown in Figure 10.42. Your name and the time required are stored in an SQLite database.

Figure 10.42 Highscore List

First, let's tackle the framework for the GUI program.

```python
import time, random, glob, sqlite3, tkinter, tkinter.messagebox as mb
def evaluation():
    ...
def start():
    ...
def action():
    ...

# Program
window = tkinter.Tk()
window.title("Mental calculation")
```

```
window.resizable(0, 0)

lbName = tkinter.Label(window, text="Your name:")
lbName.grid(row=0, column=0, sticky="w", padx=5, pady=10)
etName = tkinter.Entry(window, width=15)
etName.grid(row=0, column=2, pady=10)
buAction = tkinter.Button(window, text="Start", width=10, command=action)
buAction.grid(row=1, column=0, columnspan=3, pady=10)

taskList = []
entryList = []
for i in range(5):
    lbTemp = tkinter.Label(window, text=f"Task {i+1}:")
    lbTemp.grid(row=i+2, column=0, padx=5, pady=5)
    lbTemp = tkinter.Label(window, width=10, pady=5)
    lbTemp.grid(row=i+2, column=1, pady=5)
    taskList.append(lbTemp)
    etTemp = tkinter.Entry(window, width=15)
    etTemp.grid(row=i+2, column=2, padx=10, pady=5)
    entryList.append(etTemp)

window.mainloop()
```

Listing 10.26 game_gui.py File, Frame of the Program

We'll take a closer look at the three functions next. A label, the input field for the name, and the button appear first.

The button is initially labeled **Start**. Not until a name has been entered and the button clicked will it be labeled **Stop**. The status of this button controls whether the tasks appear or the evaluation takes place after the button is clicked.

Two empty lists are then created, which contain the references for the input and output fields after the for loop has been run. In the for loop, the following actions occur with each run:

- A label with the task number is created. The reference to the label is only required temporarily for creation and placement.
- An empty label is created in which the task will later appear. The reference to this label is also required only temporarily. The reference is assigned to the taskList list, which contains the references to the five empty labels after the for loop.
- An input field is created in which the solution determined from the player is to be entered. The reference to the input field is also only required temporarily. It is assigned to the inputList list, which contains the references to all five input fields after the for loop.

355

The action() function is called after the button has been clicked. Different actions are carried out depending on the label.

```python
def action():
    if buAction["text"] == "Start":
        if etName.get() == "":
            mb.showinfo("Name", "Please enter a name")
        else:
            start()
    else:
        evaluation()
```

Listing 10.27 game_gui.py File, action() Function

If the button is labeled **Start**, the system checks whether a name has been entered. If so, the start() function is called; otherwise, an error message appears. If the button does not have the label **Start**, the evaluation() function will be called.

Five tasks are created and displayed in the start() function, defined next.

```python
def start():
    global resultList, starttime
    resultList = []
    for i in range(5):
        a = random.randint(10,30)
        b = random.randint(10,30)
        resultList.append(a + b)
        taskList[i]["text"] = f"{a} + {b} ="
    starttime = time.time()
    buAction["text"] = "Stop"
```

Listing 10.28 game_gui.py File, start() Function

Another empty list is created that contains the results after the for loop has been run. Two random values for the task are determined in the for loop. The correct result is calculated and added to the list. The task is displayed, the start time determined, and the button label changes to **Stop**.

The inputs are evaluated in the evaluation() function. If all five tasks are solved correctly, the name and time required for solving them are entered into the SQLite database.

```python
def evaluation():
    global resultList, starttime

    difference = time.time() - starttime
```

```
correct = 0
for i in range(5):
    try:
        if int(entryList[i].get()) == resultList[i]:
            correct += 1
    except:
        pass

if correct < 5:
    mb.showinfo("No highscore", "Sorry, no highscore")
    window.destroy()
    return

if not glob.glob("game_gui_highscore.db"):
    con = sqlite3.connect("game_gui_highscore.db")
    cursor = con.cursor()
    sql = "CREATE TABLE data(name TEXT, time REAL)"
    cursor.execute(sql)
    con.close()

con = sqlite3.connect("game_gui_highscore.db")
cursor = con.cursor()
sql = f"INSERT INTO data VALUES(?,{difference})"
cursor.execute(sql, (etName.get(), ))
con.commit()
con.close()

con = sqlite3.connect("game_gui_highscore.db")
cursor = con.cursor()
sql = "SELECT * FROM data ORDER BY time LIMIT 10"
cursor.execute(sql)
output = ""
i = 1
for dr in cursor:
    output += f"{i}. {dr[0]} {round(dr[1],2)} sec.\n"
    i += 1
mb.showinfo("Highscore", output)
con.close()

window.destroy()
```

Listing 10.29 game_gui.py File, evaluation() Function

The first step is to determine the time required. The five entries are then compared with the five correct results. If not all results are correct, a message window appears to a corresponding message, and the program ends.

If all results are correct, the following happens: If no highscore database exists yet, a new one will be created. A data record with the name and time is entered in the database.

Using a SELECT statement and the LIMIT keyword, the ten data records with the shortest times are determined and displayed in a message window, rounded to two decimal places. The program then ends.

10.8   Exercise

u_gui_language

Use the *tkinter* interface in the *u_gui_language.py* file to create a simple language learning program based on an SQLite database. The logic is the same as in the u_sqlite_language exercise from Chapter 9, Section 9.2.9. Figure 10.43 shows the GUI after calling the program. Notice that the **Check** button is deactivated.

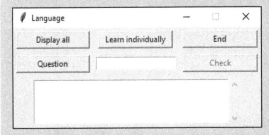

Figure 10.43 GUI after the Call

After clicking the **Display all** button, all data records are displayed in a list box, as shown in Figure 10.44.

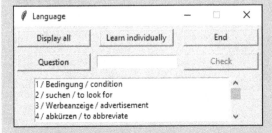

Figure 10.44 Displaying All Data Records in a List Box

After clicking the **Learn individually** button, a randomly selected data record is displayed in a message box to help you learn the term, as shown in Figure 10.45.

Figure 10.45 Displaying a Randomly Selected Data Record

After clicking the **Question** button, a randomly selected German word is displayed in a message box, as shown in Figure 10.46.

Figure 10.46 Display of a Randomly Selected Word

After clicking the **OK** button, the **Question** button is automatically deactivated, while the **Check** button is automatically activated. The English translation should now be entered in the input field.

After clicking the **Check** button, one of two possible answers appears in a message box, as shown in Figure 10.47 and Figure 10.48. The **Question** button is automatically reactivated, whereas the **Check** button is automatically deactivated.

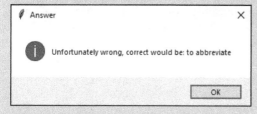

Figure 10.47 Reaction to Incorrect Answer

Figure 10.48 Reaction to Correct Answer

If the German word has been translated correctly, it will be removed from the list. If all words have been translated correctly, the list is empty, and the next time the **Question** button is clicked, the message box shown in Figure 10.49 will be displayed. The program can be terminated.

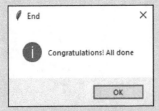

Figure 10.49 Everything Translated Correctly

If the player terminates the program before all words have been translated correctly, the query shown in Figure 10.50 will be displayed.

Figure 10.50 Exit Anyway?

After clicking the **Cancel** button, the termination process will be canceled, and the test can continue.

If you're not yet comfortable with the development of this program, you'll find some tips in Appendix A, Section A.5.

In addition to many other expansion options, the **Display all** and **Learn individually** buttons could be deactivated during a test and reactivated later.

Chapter 11
User Interfaces with PyQt

Like the *Tk* library, the *Qt* library enables the standardized programming of a *graphical user interface (GUI)* for different operating systems. For Python, *PyQt* is a freely available interface to Qt. In this chapter, I'll provide an introduction to the current version, PyQt 6.

You can install the interface with the following commands (see also Appendix A, Section A.1):

- On Windows: `pip install PyQt6`
- On Ubuntu Linux: `sudo apt install python3-pyqt6`
- On macOS: `pip install PyQt5`

Note the different notation on Ubuntu Linux. Also, on macOS, you can currently (as of spring 2024) only use the previous version, but this version is sufficient for most of the programs in this chapter. For this reason, if necessary, the `import` statement in the first program line must be adapted from `PyQt6` to `PyQt5`.

11.1 Introduction

When you use the *PyQt* interface, the creation of a GUI application comprises the following steps:

1. Creating a custom class, which inherits from the `QWidget` class, for the user interface
2. Defining a constructor of this class in which the elements of the application's user interface are created
3. Creating an object of the `QApplication` class for the application
4. Creating an object of the custom class for the user interface
5. Starting the application

11.1.1 Our First GUI Program

Our first program creates an application with a PyQt GUI that contains a single widget, namely, the **End** button, as shown in Figure 11.1. After clicking this button, the GUI is closed, and the application terminates.

Figure 11.1 Our First Program with PyQt

Let's explore our first program in depth next.

```python
import PyQt6.QtWidgets as wi
import sys

class Window(wi.QWidget):
    def __init__(self):
        super().__init__()

        buEnd = wi.QPushButton("End", self)
        buEnd.move(40, 20)
        instance = wi.QApplication.instance()
        buEnd.clicked.connect(instance.quit)

        self.setWindowTitle("...")
        self.show()

application = wi.QApplication(sys.argv)
window = Window()
sys.exit(application.exec())
```

Listing 11.1 pyqt_first.py File

The PyQt6.QtWidgets module contains the various classes for the widgets (i.e., the elements of the application window). In our example, the module is imported and given the alias wi. The sys module is required to start and end the application.

An object of the QApplication class is created in the main program to control the GUI application. The parameters of the command line are passed on as parameters for the constructor.

Next, an object of the custom Window class is created, which is used to design the application window. This class inherits from the QWidget class, the base class of the various widget classes.

The exec() method is called for the object of the QApplication class. This method starts an infinite loop in which the system waits for events. If the application window is closed, the application is also closed using the exit() method.

In the constructor of the Window class, the constructor of the QWidget base class must first be called. The elements of the application window are then created.

An object of the QPushButton class corresponds to a standard button. In our example, the button is labeled **End** and is assigned to the window object using self. The move() method moves the button from the default position 0, 0 to the desired position, measured from the top-left corner of the window.

Signals and *slots* are used to edit events in Qt. When an event occurs, in this case, the clicked event of the button, a signal is sent. The connect() method is used to connect this signal to a slot, which is a method by default. Here, this slot is the quit() method of the application instance. A reference to the instance of the application is requested beforehand using the instance() method of the QApplication class. Clicking the button exits the infinite loop, closes the GUI, and ends the application.

The setWindowTitle() method displays a text in the title bar. The show() method displays the application window.

11.1.2 Setup and Arrangement

PyQt provides the ability conveniently arrange widgets in the application window in a *grid layout* (i.e., a grid of rows and columns). In the next example, we arrange two buttons and a label in a grid layout.

```python
import PyQt6.QtWidgets as wi
import sys

class Window(wi.QWidget):
    def __init__(self):
        super().__init__()
        gr = wi.QGridLayout()

        buHello = wi.QPushButton("Hello")
        buHello.clicked.connect(self.hello)
        gr.addWidget(buHello, 0, 0)

        self.lbOutput = wi.QLabel("(empty)")
        gr.addWidget(self.lbOutput, 1, 0)

        buEnd = wi.QPushButton("End")
        buEnd.clicked.connect(wi.QApplication.instance().quit)
        gr.addWidget(buEnd, 2, 1)

        self.setWindowTitle("...")
```

11

```
        self.setLayout(gr)
        self.show()

    def hello(self):
        self.lbOutput.setText("Hello")

application = wi.QApplication(sys.argv)
window = Window()
sys.exit(application.exec())
```

Listing 11.2 pyqt_arrangement.py File

After clicking the **Hello** button, the text "Hello" appears on the label, as shown in Figure 11.2.

Figure 11.2 Arrangement of the Widgets in a Grid

The grid is created using an object of the QGridLayout class. The individual widgets are no longer assigned to the window but to the grid. The assignment is made by calling the addWidget() method for the QGridLayout object. A reference to the respective widget is specified as a parameter, followed by the numbers for the row and column of the grid. The numbering starts at 0.

When the buHello button is clicked, the hello() method is called for the object of the Window class. The button is displayed with the parameters 0, 0 in the top-left cell of the grid.

Texts or images can be displayed in a label, which can be created using the QLabel class. The content of the label is to be changed in the hello() method. The reference to the label is therefore declared as a property of the Window class using self. Otherwise, the reference would only be a local variable of the __init__() method, which is not known in other methods.

The layout of the GUI is assigned by calling the setLayout() method for the object of the Window class.

In the hello() method, the setText() method is called for the label object to change the text.

11.1.3 Size of the Application Window

The size of the application window depends on the sizes of the elements it contains. The size of a label depends on the content and, in the case of text content, on the font size. The various operating systems use different default font sizes, which is why there may be small differences in the size of the application window for the same PyQt program, when compared on different operating systems.

You could fix the size of the application window using the setFixedSize() method. However, several disadvantages would arise for the reasons mentioned earlier, namely, the following:

- If text that is too long is displayed in a label while the program is running, the text will be truncated.

- If the texts are displayed with a larger font by default on another operating system, they may be truncated when the program is started.

11.1.4 Event Methods for Multiple Widgets

You can use a *lambda function* (also known as *anonymous functions*; see Chapter 5, Section 5.6.8) to call the same function from multiple widgets. In this process, you can also pass parameters, which is impossible without an anonymous function.

Let's look at an example with two buttons that call the same function.

```python
import PyQt6.QtWidgets as wi
import sys

class Window(wi.QWidget):
    def __init__(self):
        super().__init__()
        gr = wi.QGridLayout()

        buOne = wi.QPushButton("1")
        buOne.clicked.connect(lambda:self.out(1))
        gr.addWidget(buOne, 0, 0)

        buTwo = wi.QPushButton("2")
        buTwo.clicked.connect(lambda:self.out(2))
        gr.addWidget(buTwo, 1, 0)

        self.lbOutput = wi.QLabel()
        gr.addWidget(self.lbOutput, 2, 0)

        self.setWindowTitle("...")
        self.setLayout(gr)
```

11

```
        self.show()

    def out(self, p):
        self.lbOutput.setText(str(p))

application = wi.QApplication(sys.argv)
window = Window()
sys.exit(application.exec())
```

Listing 11.3 pyqt_lambda.py File

After clicking one of the two buttons, an anonymous function is called. The out() function is called with the parameter 1 or 2. The value of the parameter then appears in the
label, as shown in Figure 11.3.

Figure 11.3 After Clicking the 1 Button

11.2   Widget Types

In this section, I'll introduce you to many other widgets, including QLineEdit, QTextEdit,
QListWidget, QComboBox, QSpinBox, QRadioButton, QButtonGroup, QCheckBox, and QSlider.
The QLabel widget is also used to display images, colors, and hyperlinks.

11.2.1   Single-Line Input Field

The content of many widgets can be evaluated both during operation and afterwards.

An object of the QLineEdit class serves as a single-line input field. The textChanged signal can be used to respond immediately when the content of the input field is changed,
i.e., when individual characters are added or removed. However, the content of the
input field is usually processed at a later stage, for example after a button has been
clicked.

Both events are used in our next program.

```
import PyQt6.QtWidgets as wi
import sys

class Window(wi.QWidget):
    def __init__(self):
```

```
        super().__init__()
        gr = wi.QGridLayout()

        lbInput = wi.QLabel("Your input:")
        gr.addWidget(lbInput, 0, 0)

        self.leInput = wi.QLineEdit()
        self.leInput.textChanged.connect(self.input)
        gr.addWidget(self.leInput, 1, 0)

        buDouble = wi.QPushButton("Double")
        buDouble.clicked.connect(self.double)
        gr.addWidget(buDouble, 2, 0)

        self.lbOutput = wi.QLabel("(empty)")
        gr.addWidget(self.lbOutput, 3, 0)

        buEnd = wi.QPushButton("End")
        buEnd.clicked.connect(wi.QApplication.instance().quit)
        gr.addWidget(buEnd, 4, 1)

        self.setWindowTitle("Input")
        self.setLayout(gr)
        self.show()

    def input(self, tx):
        self.lbOutput.setText(f"Input: {tx}")

    def double(self):
        try:
            number = float(self.leInput.text())
            self.lbOutput.setText(f"{number * 2}")
        except:
            self.lbOutput.setText("No number")

application = wi.QApplication(sys.argv)
window = Window()
sys.exit(application.exec())
```

Listing 11.4 pyqt_input.py File

The leInput reference refers to the QLineEdit object. The textChanged signal of this object is linked to the input() method. The current content of the input field is automatically transferred to this method. Here, this content is output in a label, as shown in Figure 11.4.

Clicking the **Double** button is linked to the double() method. In this method, the content of the input field is determined using the text() method of the QLineEdit object. If this content can be converted into a valid number using the float() function, this number is doubled and output in a label, as shown in Figure 11.5. Otherwise, an error message will appear.

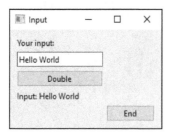

Figure 11.4 The Change of Content Event

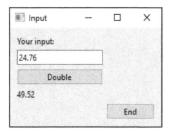

Figure 11.5 Clicking the Double Button

11.2.2   Hidden Inputs and Disabling Widgets

A QLineEdit object also enables hidden input, for example, for a password. Widgets can be enabled or disabled to simplify operations. In our next program, which involves hidden input for a password, a widget is both enabled and disabled.

```python
import PyQt6.QtWidgets as wi
import sys

class Window(wi.QWidget):
    def __init__(self):
        super().__init__()
        gr = wi.QGridLayout()

        lbPassword = wi.QLabel("Your password:")
        gr.addWidget(lbPassword, 0, 0)
```

```
        self.lePassword = wi.QLineEdit()
        self.lePassword.setEchoMode(wi.QLineEdit.EchoMode.Password)
        gr.addWidget(self.lePassword, 1, 0)

        buCheck = wi.QPushButton("Check")
        buCheck.clicked.connect(self.check)
        gr.addWidget(buCheck, 2, 0)

        self.lbOutput = wi.QLabel("(empty)")
        gr.addWidget(self.lbOutput, 3, 0)

        self.buEnd = wi.QPushButton("End")
        self.buEnd.setEnabled(False)
        self.buEnd.clicked.connect(wi.QApplication.instance().quit)
        gr.addWidget(self.buEnd, 4, 1)

        self.setWindowTitle("Password")
        self.setLayout(gr)
        self.show()

    def check(self):
        pw = self.lePassword.text()
        if pw == "Bingo":
            self.lbOutput.setText("Access allowed")
        else:
            self.lbOutput.setText("Access not allowed")
        self.lePassword.setText("")
        self.buEnd.setEnabled(True)

application = wi.QApplication(sys.argv)
window = Window()
sys.exit(application.exec())
```

Listing 11.5 pyqt_password.py File

The lePassword reference refers to a QLineEdit object. The setEchoMode() method is called for this object. An element of the EchoMode enumeration of the QLineEdit class must be transferred as a parameter. If the Password element is used, a bullet character appears for each character entered, as shown in Figure 11.6.

The **End** button was initially disabled by calling the setEnabled() method with the parameter False and cannot be operated.

Figure 11.6 Hidden Input

By clicking the **Check** button, the check() method is called. The content of the input field is determined. Depending on the content, one of two different messages appears. After that, the content is removed. The **End** button is enabled by calling the setEnabled() method with the True parameter and can now be clicked, as shown in Figure 11.7.

Figure 11.7 Enabling a Widget

11.2.3   Multiline Input Field

An object of the QTextEdit class corresponds to a multiline input field for displaying and editing large amounts of data. With the help of Hypertext Markup Language (HTML), you can access a wide range of formatting options and use the multiline input field as a *what you see is what you get (WYSIWYG)* editor. If the size of the widget is not sufficient, a vertical scrollbar will be displayed.

In our next example, we can only show a small selection of the available capabilities. A QTextEdit object displays data loaded from a text file. You can also save the content of the input field in a text file.

```
import PyQt6.QtWidgets as wi
import sys

class Window(wi.QWidget):
    def __init__(self):
        super().__init__()
        gr = wi.QGridLayout()
```

```
        lbText = wi.QLabel("Text:")
        gr.addWidget(lbText, 0, 0)

        self.teText = wi.QTextEdit()
        self.teText.setFixedHeight(80)
        gr.addWidget(self.teText, 1, 0, 1, 3)

        buLoad = wi.QPushButton("Load from file")
        buLoad.clicked.connect(self.load)
        gr.addWidget(buLoad, 2, 0)

        buSave = wi.QPushButton("Save to file")
        buSave.clicked.connect(self.save)
        gr.addWidget(buSave, 2, 1)

        buDelete = wi.QPushButton("Delete text")
        buDelete.clicked.connect(self.delete)
        gr.addWidget(buDelete, 2, 2)

        self.lbOutput = wi.QLabel("(empty)")
        gr.addWidget(self.lbOutput, 3, 0)

        self.buEnd = wi.QPushButton("End")
        self.buEnd.clicked.connect(wi.QApplication.instance().quit)
        gr.addWidget(self.buEnd, 3, 3)

        self.setWindowTitle("TextEdit")
        self.setLayout(gr)
        self.show()

    def load(self):
        self.teText.setText("")
        try:
            f = open("pyqt_multiline.txt")
            self.teText.setText(f.read())
            f.close()
        except:
            self.teText.setText("File not opened")

    def save(self):
        try:
            f = open("pyqt_multiline.txt", "w")
            f.write(self.teText.toPlainText())
            f.close()
```

```
        except:
            self.lbOutput.setText("File not opened")

    def delete(self):
        self.teText.setText("")

application = wi.QApplication(sys.argv)
window = Window()
sys.exit(application.exec())
```

Listing 11.6 pyqt_multiline.py File

The teText reference refers to a QTextEdit object, as shown in Figure 11.8. The height of a widget can be fixed using the setFixedHeight() method.

This widget is supposed extend over three columns within the grid. For this purpose, two additional parameters with the values 1 (row) and 3 (columns) are appended when the addWidget() method is called. The two optional parameters both have the default values 1 and 1.

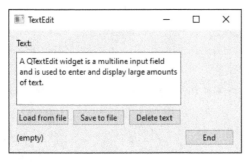

Figure 11.8 A Multiline Input Field

After clicking the **Load from file** button, the load() method is called. In this method, the text field is emptied using the setText() method. Then, a text file is opened. Its content is read in full and displayed in the text field. If an error occurs, an error message appears in the text field instead.

After clicking the **Save to file** button, the save() method is called. In this method, the pure text content of the text field is determined using the toPlainText() method and written to the file. If an error occurs, an error message is also displayed. To avoid having the message overwrite the content of the text field, it appears in the label below the text field. The toHtml() method would determine the HTML content of a text field.

The **Delete text** button calls the delete() method in which the text field is deleted. The **End** button is located in column 3 of the grid, since the three other buttons are located in columns 0, 1, and 2, and the text field extends across these three columns.

11.2.4   List with a Single Selection Option

You can use an object of the QListWidget class to make a selection from a list. This section deals with the selection of a single element, while the next section deals with the simultaneous selection of multiple elements.

The QListWidget object can react immediately to a change of selection. You can also evaluate your selection afterwards. Both options are shown in our next example.

```python
import PyQt6.QtWidgets as wi
import sys

class Window(wi.QWidget):
    def __init__(self):
        super().__init__()
        gr = wi.QGridLayout()

        lbSelection = wi.QLabel("Your selection:")
        gr.addWidget(lbSelection, 0, 0)

        self.liSelection = wi.QListWidget()
        self.liSelection.setFixedHeight(80)
        city = ["Houston", "Seattle", "Boston", "Detroit", "Tampa",
            "Denver", "Portland", "Hartford", "Fremont", "Atlanta"]
        for c in city:
            self.liSelection.addItem(c)
        self.liSelection.setCurrentRow(0)
        self.liSelection.itemClicked.connect(self.selection)
        gr.addWidget(self.liSelection, 1, 0)

        buOutput = wi.QPushButton("Output")
        buOutput.clicked.connect(self.output)
        gr.addWidget(buOutput, 2, 0)

        self.lbOutput = wi.QLabel("(empty)")
        gr.addWidget(self.lbOutput, 3, 0)

        buEnd = wi.QPushButton("End")
        buEnd.clicked.connect(wi.QApplication.instance().quit)
        gr.addWidget(buEnd, 4, 1)

        self.setWindowTitle("ListWidget")
        self.setLayout(gr)
        self.show()
```

11

```
    def selection(self, it):
        self.lbOutput.setText(f"Selection: {it.text()}")

    def output(self):
        it = self.liSelection.currentItem()
        self.lbOutput.setText(f"Output: {it.text()}")

application = wi.QApplication(sys.argv)
window = Window()
sys.exit(application.exec())
```

Listing 11.7 pyqt_list.py File

The liSelection reference refers to a QListWidget object. If not all entries are visible at the same time, the object is given a vertical scrollbar, as shown in Figure 11.9.

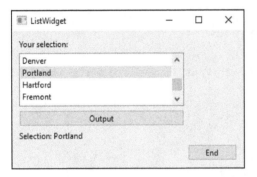

Figure 11.9 A List with a Single Selection Option

The displayed entries are of the QListWidgetItem type. These entries are added to the QListWidget using the addItem() method, in this case, from a list. The displayed text of the entry is transferred as a parameter.

The setCurrentRow() method is used to highlight an entry so that there is also a selection if the QListWidget object has not been operated. The index of the entry is passed as a parameter. The first entry has the index 0.

The selection of an entry corresponds to the itemClicked signal. This signal is connected with the select() method. A reference to the selected QListWidgetItem object is automatically transferred to this method. The text() method of the object returns the displayed text of the entry.

After clicking the **Output** button, the output() method is called. In this method, the currently selected item is determined using the currentItem() method. The corresponding text is output.

11.2.5   List with Multiple Selection Options

The setSelectionMode() method is used to set the selection mode for a QListWidget object. Multiple selections are possible in the next program. Pressing Ctrl, you can select multiple entries, which do not have to be consecutive. In addition, you can use the Shift key to select multiple consecutive entries. The current selection appears in a second QListWidget object, as shown in Figure 11.10.

```python
import PyQt6.QtWidgets as wi
import sys

class Window(wi.QWidget):
    def __init__(self):
        super().__init__()
        gr = wi.QGridLayout()

        lbSelection = wi.QLabel("Your selection:")
        gr.addWidget(lbSelection, 0, 0)

        self.liSelection = wi.QListWidget()
        self.liSelection.setFixedHeight(80)
        self.liSelection.setSelectionMode(
            wi.QAbstractItemView.SelectionMode.ExtendedSelection)
        city = ["Houston", "Seattle", "Boston", "Detroit", "Tampa",
            "Denver", "Portland", "Hartford", "Fremont", "Atlanta"]
        for c in city:
            self.liSelection.addItem(c)
        self.liSelection.setCurrentRow(0)
        self.liSelection.itemClicked.connect(self.output)
        gr.addWidget(self.liSelection, 1, 0)

        buOutput = wi.QPushButton("Output")
        buOutput.clicked.connect(self.output)
        gr.addWidget(buOutput, 2, 0)

        self.liOutput = wi.QListWidget()
        self.liOutput.setFixedHeight(80)
        gr.addWidget(self.liOutput, 3, 0)

        buEnd = wi.QPushButton("End")
        buEnd.clicked.connect(wi.QApplication.instance().quit)
        gr.addWidget(buEnd, 4, 1)

        self.setWindowTitle("ListWidget")
```

11

```
        self.setLayout(gr)
        self.show()

    def output(self):
        while self.liOutput.count() > 0:
            self.liOutput.takeItem(0)
        for it in self.liOutput.selectedItems():
            self.liOutput.addItem(it.text())

application = wi.QApplication(sys.argv)
window = Window()
sys.exit(application.exec())
```

Listing 11.8 pyqt_list_multiple.py File

An element of the SelectionMode enumeration from the QAbstractItemView class is expected as a parameter of the setSelectionMode() method. This class serves as an abstract base class for various types of widgets that have entries.

Both changing the selection or clicking the **Output** button lead to the output() method. The second QListWidget object is emptied and then filled with the selected entries. The count() method returns the number of entries. The takeItem() method removes the item with the specified index. In this scenario, the first entry is removed multiple times using a while loop until the list is empty.

The selectedItems() method returns a list of QListWidgetItem objects that are listed in the order in which they were selected. The list is run through using a loop. The text of each item is added to the second QListWidget object using the addItem() method, as shown in Figure 11.10.

Figure 11.10 A List with Multiple Selection Options

11.2.6   Combo Boxes

You can use an object of the QComboBox class to create a combo box. You can select entries from a list or enter text as in a text field. The "Selection changed" and "Text changed" events are noted for the object, as illustrated in our next program.

```python
import PyQt6.QtWidgets as wi
import sys

class Window(wi.QWidget):
    def __init__(self):
        super().__init__()
        gr = wi.QGridLayout()

        self.lbResult = wi.QLabel("Your selection or input:")
        gr.addWidget(self.lbResult, 0, 0)

        self.cbSelection = wi.QComboBox()
        self.cbSelection.setEditable(True)
        city = ["Houston", "Seattle", "Boston", "Detroit", "Tampa",
            "Denver", "Portland", "Hartford", "Fremont", "Atlanta"]
        for c in city:
            self.cbSelection.addItem(c)
        self.cbSelection.currentTextChanged.connect(self.input)
        self.cbSelection.currentIndexChanged.connect(self.selection)
        gr.addWidget(self.cbSelection, 1, 0)

        buOutput = wi.QPushButton("Output")
        buOutput.clicked.connect(self.output)
        gr.addWidget(buOutput, 2, 0)

        buEnd = wi.QPushButton("End")
        buEnd.clicked.connect(wi.QApplication.instance().quit)
        gr.addWidget(buEnd, 3, 1)

        self.setWindowTitle("ComboBox")
        self.setLayout(gr)
        self.show()

    def selection(self, ix):
        tx = self.cbSelection.currentText()
        self.lbResult.setText(f"Selection: {tx}, Index: {ix}")
```

```
def input(self, tx):
    self.lbResult.setText(f"Input: {tx}")

def output(self):
    tx = self.cbSelection.currentText()
    ix = self.cbSelection.currentIndex()
    self.lbResult.setText(f"Output: {tx}, Index: {ix}")

application = wi.QApplication(sys.argv)
window = Window()
sys.exit(application.exec())
```

Listing 11.9 pyqt_combo.py File

By default, no text can be entered in the QComboBox object. You can change this setting using the setEditable() method, with the True parameter. As with the QListWidget object, the entries are added using the addItem() method.

If text is entered or changed, the currentTextChanged signal will be set. This signal leads to the input() method, which automatically receives the current text in the QComboBox object. The text then is output, as shown in Figure 11.11.

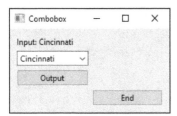

Figure 11.11 Entering a Text

When the selection changes to another entry, the signal currentIndexChanged is set. This signal leads to the select() method, which automatically receives the index of the current entry. The text of the current entry can be determined using the currentText() method. Both pieces of information are output, as shown in Figure 11.12.

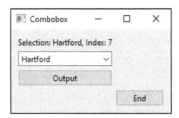

Figure 11.12 Selecting an Entry

After clicking the **Output** button, the `output()` method is called. The `currentIndex()` method returns the index of the current entry.

11.2.7   Spin Boxes

A spin box allows you to conveniently select or enter a numerical value from a limited numerical range. Values outside the numerical range can neither be selected nor entered. Objects of the `QSpinBox` class are integers, while objects of the `QDoubleSpinBox` class are numbers with decimal places.

We'll work with objects of both classes in the following program.

```
import PyQt6.QtWidgets as wi
import sys

class Window(wi.QWidget):
    def __init__(self):
        super().__init__()
        gr = wi.QGridLayout()

        self.lbResult = wi.QLabel("Your selection or input:")
        self.lbResult.setFixedWidth(150)
        gr.addWidget(self.lbResult, 0, 0)

        self.spSelectionInt = wi.QSpinBox()
        self.spSelectionInt.setMinimum(10)
        self.spSelectionInt.setMaximum(30)
        self.spSelectionInt.setSingleStep(2)
        self.spSelectionInt.setValue(16)
        self.spSelectionInt.valueChanged.connect(self.change)
        gr.addWidget(self.spSelectionInt, 1, 0)

        self.spSelectionDbl = wi.QDoubleSpinBox()
        self.spSelectionDbl.setMinimum(2.8)
        self.spSelectionDbl.setMaximum(7.2)
        self.spSelectionDbl.setSingleStep(0.2)
        self.spSelectionDbl.setValue(3.4)
        self.spSelectionDbl.setDecimals(1)
        self.spSelectionDbl.valueChanged.connect(self.change)
        gr.addWidget(self.spSelectionDbl, 2, 0)

        buOutput = wi.QPushButton("Output")
        buOutput.clicked.connect(self.output)
        gr.addWidget(buOutput, 3, 0)
```

11

```
            buEnd = wi.QPushButton("End")
            buEnd.clicked.connect(wi.QApplication.instance().quit)
            gr.addWidget(buEnd, 4, 1)

            self.setWindowTitle("Spinbox")
            self.setLayout(gr)
            self.show()

    def change(self, value):
        if self.sender() is self.spSelectionInt:
            self.lbResult.setText(f"Selection: {value}")
        else:
            self.lbResult.setText(f"Selection: {round(value,1)}")

    def output(self):
        valueInt = self.spSelectionInt.value()
        valueDbl = round(self.spSelectionDbl.value(), 1)
        self.lbResult.setText(f"Output: {valueInt} and {valueDbl}")

application = wi.QApplication(sys.argv)
window = Window()
sys.exit(application.exec())
```

Listing 11.10 pyqt_spin.py File

The setMinimum(), setMaximum(), and setValue() methods are used to set the limits of the number range and the current value.

Using the setSingleStep() method, you can set the increment when the arrows on the spin box are clicked on. The default value is 1. You can enter a value that lies between two steps through an input, but never a value outside the number range.

In a spin box for numbers with decimal places, the setDecimals() method is used to set the number of decimal places for the input and the display in the spin box.

The value of a spin box can be changed by making an entry or a selection. Both actions activate the valueChanged signal. In this case, this signal leads to the change() method for both spin boxes. The current value of the spin box that set the signal (i.e., triggered the event) is automatically available in the method. The text then is output, as shown in Figure 11.13.

The definition of the number of decimal places only applies to the input and the display in the spin box, not to the resulting value or its display in the output field. This value must therefore also be rounded to one decimal place. Since the change() method reacts to both spin boxes, we must determine the sender of the signal.

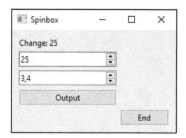

Figure 11.13 Value of the First Spin Box after Input

When an event occurs, the sender() method of the application window returns a reference to the object that set the signal. This reference is compared with the reference to
the first spin box. In this way, the output can be designed appropriately, as shown in
Figure 11.14.

Figure 11.14 Value of the Second Spin Box after Selection

After clicking the **Output** button, the output() method is called. The value() method
returns the current value of a spin box, as shown in Figure 11.15.

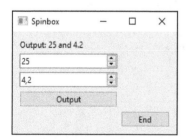

Figure 11.15 Values of the Two Spin Boxes

11.2.8   Radio Buttons

Objects of the QRadioButton class serve as radio buttons, which make it easy to choose
between several options.

In our next program, we'll use radio buttons for selecting a color.

```python
import PyQt6.QtWidgets as wi
import sys

class Window(wi.QWidget):
    def __init__(self):
        super().__init__()
        gr = wi.QGridLayout()

        self.lbSelection = wi.QLabel("Your selection:")
        gr.addWidget(self.lbSelection, 0, 0, 1, 3)

        self.rbRed = wi.QRadioButton("Red")
        self.rbRed.toggled.connect(self.selection)
        gr.addWidget(self.rbRed, 1, 0)

        self.rbYellow = wi.QRadioButton("Yellow")
        self.rbYellow.setChecked(True)
        self.rbYellow.toggled.connect(self.selection)
        gr.addWidget(self.rbYellow, 1, 1)

        self.rbBlue = wi.QRadioButton("Blue")
        self.rbBlue.toggled.connect(self.selection)
        gr.addWidget(self.rbBlue, 1, 2)

        buOutput = wi.QPushButton("Output")
        buOutput.clicked.connect(self.output)
        gr.addWidget(buOutput, 2, 0, 1, 3)

        buEnd = wi.QPushButton("End")
        buEnd.clicked.connect(wi.QApplication.instance().quit)
        gr.addWidget(buEnd, 3, 3)

        self.setWindowTitle("Radiobutton")
        self.setLayout(gr)
        self.show()

    def selection(self):
        rb = self.sender()
        if rb.isChecked():
            self.lbSelection.setText(f"Selection: {rb.text()}")

    def output(self):
        if self.rbRed.isChecked():
            rb = self.rbRed
```

```
        elif self.rbYellow.isChecked():
            rb = self.rbYellow
        else:
            rb = self.rbBlue
        self.lbSelection.setText(f"Output: {rb.text()}")

application = wi.QApplication(sys.argv)
window = Window()
sys.exit(application.exec())
```

Listing 11.11 pyqt_radio.py File

The output label and the button below the three radio buttons each extend over three columns.

At the beginning, one of the radio buttons should already be selected using the set-Checked() method so that a selection is made in every case. The toggled signal is set if a radio button changes its status. This event happens both when a radio button is selected and when the selection is lost by selecting another radio button.

In our example, the toggled signal leads to the select() method for all radio buttons. The sender() method returns a reference to the radio button for which the status has changed. The isChecked() method determines whether a radio button has been checked. In this case, the text of the radio button will be displayed, as shown in Figure 11.16. If a radio button has lost its selection, nothing will be displayed.

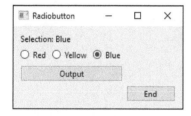

Figure 11.16 Selecting a Radio Button

After clicking the **Output** button, the output() method is called. The reference of the currently selected radio button is determined and transferred in a multiple branch.

11.2.9   Multiple Groups of Radio Buttons

Objects of the QButtonGroup class are not visible themselves but can be used to logically separate multiple groups of radio buttons. At the same time, they make it easier to determine the currently selected radio button within a group. In our next program, a color and a size can be selected in this way.

```python
import PyQt6.QtWidgets as wi
import sys

class Window(wi.QWidget):
    def __init__(self):
        super().__init__()
        gr = wi.QGridLayout()

        self.lbSelection = wi.QLabel("Your selection:")
        gr.addWidget(self.lbSelection, 0, 0, 1, 3)

        rbRed = wi.QRadioButton("Red")
        gr.addWidget(rbRed, 1, 0)

        rbYellow = wi.QRadioButton("Yellow")
        rbYellow.setChecked(True)
        gr.addWidget(rbYellow, 1, 1)

        rbBlue = wi.QRadioButton("Blue")
        gr.addWidget(rbBlue, 1, 2)

        self.gpColor = wi.QButtonGroup()
        self.gpColor.addButton(rbRed)
        self.gpColor.addButton(rbYellow)
        self.gpColor.addButton(rbBlue)
        self.gpColor.buttonClicked.connect(self.selection)

        rbSmall = wi.QRadioButton("small")
        gr.addWidget(rbSmall, 2, 0)

        rbMedium = wi.QRadioButton("medium")
        gr.addWidget(rbMedium, 2, 1)

        rbLarge = wi.QRadioButton("large")
        rbLarge.setChecked(True)
        gr.addWidget(rbLarge, 2, 2)

        self.gpSize = wi.QButtonGroup()
        self.gpSize.addButton(rbSmall)
        self.gpSize.addButton(rbMedium)
        self.gpSize.addButton(rbLarge)
        self.gpSize.buttonClicked.connect(self.selection)
```

```
        buOutput = wi.QPushButton("Output")
        buOutput.clicked.connect(self.output)
        gr.addWidget(buOutput, 3, 0, 1, 3)

        buEnd = wi.QPushButton("End")
        buEnd.clicked.connect(wi.QApplication.instance().quit)
        gr.addWidget(buEnd, 4, 3)

        self.setWindowTitle("Group")
        self.setLayout(gr)
        self.show()

    def output(self):
        rbColor = self.gpColor.checkedButton()
        rbSize = self.gpSize.checkedButton()
        self.lbSelection.setText(
            f"Output: {rbColor.text()} and {rbSize.text()}")

    def selection(self, rb):
        self.lbSelection.setText(f"Selection: {rb.text()}")

application = wi.QApplication(sys.argv)
window = Window()
sys.exit(application.exec())
```

Listing 11.12 pyqt_group.py File

This time, the references to the radio buttons do not have to be properties of the class because the information is determined via the groups.

The gpColor and gpSize references refer to the two QButtonGroup objects. A radio button is assigned to a group by calling the addButton() method for the group and transferring the reference to the relevant radio button.

The buttonClicked signal is triggered for a group if one of the radio buttons in this group is selected. The signal is connected to the select() method. When the method is called, the reference to the selected radio button is automatically made available. The corresponding text is output.

After clicking the **Output** button, the output() method is called. The checkedButton() method returns a reference to the selected radio button for each group. The texts of the selected buttons of the two groups are output, as shown in Figure 11.17.

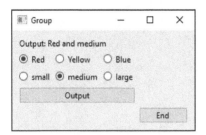

Figure 11.17 Multiple Groups of Radio Buttons

11.2.10   Checkboxes

A checkbox can be switched on or off. This element is used to represent a state in which there are two possibilities. If multiple checkboxes are available, they operate independently of each other.

The properties of a drawing object can be set with a frame or without a frame, and filled in with color or not filled in with color, as illustrated in our next program.

```python
import PyQt6.QtWidgets as wi
import sys

class Window(wi.QWidget):
    def __init__(self):
        super().__init__()
        gr = wi.QGridLayout()

        self.lbSelection = wi.QLabel("Your selection:")
        gr.addWidget(self.lbSelection, 0, 0, 1, 2)

        self.cbFrame = wi.QCheckBox("Frame")
        self.cbFrame.stateChanged.connect(self.change)
        gr.addWidget(self.cbFrame, 1, 0)

        self.cbFilling = wi.QCheckBox("Filling")
        self.cbFilling.setChecked(True)
        self.cbFilling.stateChanged.connect(self.change)
        gr.addWidget(self.cbFilling, 1, 1)

        buOutput = wi.QPushButton("Output")
        buOutput.clicked.connect(self.output)
        gr.addWidget(buOutput, 2, 0, 1, 2)
```

```
        buEnd = wi.QPushButton("End")
        buEnd.clicked.connect(wi.QApplication.instance().quit)
        gr.addWidget(buEnd, 3, 2)

        self.setWindowTitle("Checkbox")
        self.setLayout(gr)
        self.show()

    def change(self):
        cb = self.sender()
        tx = cb.text()
        if cb.isChecked():
            self.lbSelection.setText(f"{tx}: Yes")
        else:
            self.lbSelection.setText(f"{tx}: No")

    def output(self):
        if self.cbFrame.isChecked():
            tx = "Frame: Yes, "
        else:
            tx = "Frame: No, "
        if self.cbFilling.isChecked():
            tx += "Filling: Yes"
        else:
            tx += "Filling: No"
        self.lbSelection.setText(tx)

application = wi.QApplication(sys.argv)
window = Window()
sys.exit(application.exec())
```

Listing 11.13 pyqt_check.py File

The setChecked() method is used to set the checkbox via program code. The checkbox is switched on with the True parameter and switched off with the False parameter. The stateChanged signal is set when the state changes. In this case, the signal for both objects of the QCheckBox class leads to the change() method.

This method first determines which widget has sent the signal. The isChecked() method is then used to check whether the checkbox is activated, and a corresponding output is generated, as shown in Figure 11.18.

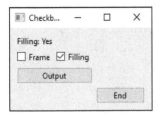

Figure 11.18 After Activating the Filling Checkbox

After clicking the **Output** button, the output() method is called. The status of the two checkboxes is determined and output, as shown in Figure 11.19.

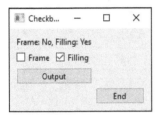

Figure 11.19 Output of the Status of Both Checkboxes

11.2.11   Sliders

You can use a QSlider object to conveniently set an integer value within a specific number range and display it clearly at the same time. A slider can be arranged vertically or horizontally and can be operated using the mouse, the arrow keys, or the [PageUp] and [PageDown] keys. For a better display, you can add a bar with markers to the slider (called *tickmarks*).

These elements are illustrated in the next program.

```python
import PyQt6.QtWidgets as wi
import PyQt6.QtCore as co
import sys

class Window(wi.QWidget):
    def __init__(self):
        super().__init__()
        gr = wi.QGridLayout()

        self.lbSelection = wi.QLabel("Value: 24")
        gr.addWidget(self.lbSelection, 0, 0)

        self.slValue = wi.QSlider()
        self.slValue.setFixedWidth(200)
```

```
        self.slValue.setOrientation(co.Qt.Orientation.Horizontal)
        self.slValue.setMinimum(10)
        self.slValue.setMaximum(50)
        self.slValue.setValue(24)
        self.slValue.setSingleStep(2)
        self.slValue.setPageStep(10)
        self.slValue.setTickPosition(wi.QSlider.TickPosition.TicksBelow)
        self.slValue.setTickInterval(5)
        self.slValue.valueChanged.connect(self.change)
        gr.addWidget(self.slValue, 1, 0)

        buOutput = wi.QPushButton("Output")
        buOutput.clicked.connect(self.output)
        gr.addWidget(buOutput, 2, 0)

        buEnd = wi.QPushButton("End")
        buEnd.clicked.connect(wi.QApplication.instance().quit)
        gr.addWidget(buEnd, 3, 1)

        self.setWindowTitle("Slider")
        self.setLayout(gr)
        self.show()

    def change(self, value):
        self.lbSelection.setText(f"After change: {value}")

    def output(self):
        value = self.slValue.value()
        self.lbSelection.setText(f"Output: {value}")

application = wi.QApplication(sys.argv)
window = Window()
sys.exit(application.exec())
```

Listing 11.14 pyqt_slider.py File

By default, a slider is arranged vertically. To set the orientation, call the setOrienta-
tion() method. The Horizontal element of the Orientation enumeration from the
PyQt6.QtCore.Qt module is used as a parameter to create a horizontal arrangement.

The setFixedWidth(), setMinimum(), setMaximum(), and setValue() methods are used to
set a fixed width, the number range and the current value.

By default, the slider can be moved by using the mouse. To set the increment for small
steps that are executed using the arrow keys, you can use the setSingleStep() method.

11

The setPageStep() method is used to set the increment for large steps that are made using the PageUp and PageDown keys or by clicking on the slider.

To display the tickmarks, you must call the setTickPosition() method. The TicksBelow element from the TickPosition enumeration of the QSlider class is used to arrange the tickmarks below the slider. You can use the setTickInterval() method to set the interval between the tickmarks.

When the slider is operated, the valueChanged signal is set, which leads to the change() method in this case. This method automatically receives the current value of the slider, which is then output, as shown in Figure 11.20.

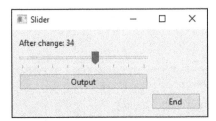

Figure 11.20 After Changing the Value

After clicking the **Output** button, the output() method is called. The current value of the slider is determined using the value() method.

11.2.12   Images, Formats, and Hyperlinks

An object of the QLabel class not only can display text, but this text can also be oriented or form a hyperlink. In addition, images can be displayed in labels as well.

The function and display of a label can be further customized by using certain HTML or *Cascading Style Sheets (CSS)* specifications. HTML and CSS provide extensive options for structuring and formatting web documents.

The following program illustrates some of these options.

```python
import PyQt6.QtWidgets as wi
import PyQt6.QtGui as gu
import PyQt6.QtCore as co
import sys

class Window(wi.QWidget):
    def __init__(self):
        super().__init__()
        gr = wi.QGridLayout()

        lbImage = wi.QLabel()
        pm = gu.QPixmap("rheinwerk.png")
```

```
        lbImage.setPixmap(pm)
        gr.addWidget(lbImage, 0, 0)

        lbColor = wi.QLabel("Text")
        lbColor.setFixedHeight(50)
        lbColor.setStyleSheet(
            "background-color:#A0A0A0; border:2px solid #000000;")
        lbColor.setAlignment(
            co.Qt.AlignmentFlag.AlignBottom | co.Qt.AlignmentFlag.AlignRight)
        gr.addWidget(lbColor, 1, 0)

        lbLink = wi.QLabel(
            "<a href='https://rheinwerk-computing.com/'>Rheinwerk Publishing</a>")
        lbLink.setOpenExternalLinks(True)
        gr.addWidget(lbLink, 2, 0)

        buEnd = wi.QPushButton("End")
        buEnd.clicked.connect(wi.QApplication.instance().quit)
        gr.addWidget(buEnd, 3, 0)

        self.setWindowTitle("Label")
        self.setLayout(gr)
        self.show()

application = wi.QApplication(sys.argv)
window = Window()
sys.exit(application.exec())
```

Listing 11.15 pyqt_label.py File

An object of the QPixmap class from the PyQt6.QtGui module represents an image from an image file. The setPixmap() method can be called for a QLabel object. A reference to a QPixmap object is transferred to the method. The represented image is displayed in the label, as shown in Figure 11.21.

The setStyleSheet() method enables you to format a label. The background-color CSS property is used to specify the background color, and the border CSS property is used to design the border, with values for the thickness, line type, and color.

The setAlignment() method is required to set the horizontal and vertical alignment of the text within the label. By default, the text is displayed left aligned and vertically centered. Using the AlignBottom and AlignRight elements of the AlignmentFlag enumeration from the QtCore module, the text is displayed right aligned at the bottom. The elements are bit flags that are linked together using the operator for the *bitwise OR*, that is, a pipe (|).

Figure 11.21 Images, Formats, and Hyperlinks

The text in a label can also be formatted using HTML tags. An HTML container is used in this case to create a hyperlink. In addition, the setOpenExternalLinks() method must be called with the True parameter.

11.2.13   Standard Dialog Boxes

An object of the QMessageBox class is a standard, ready-made dialog box for the transmission of messages. You can use this kind of dialog box to transmit information, issue a warning, report an error, and request specific answers. In the next program, various dialog boxes are each called via a separate button. If the operation of the dialog box produces an answer, then this answer is displayed in a label in the application window.

Let's consider the first part of our program with GUI elements.

```
import PyQt6.QtWidgets as wi
import sys

class Window(wi.QWidget):
    def __init__(self):
        super().__init__()
        gr = wi.QGridLayout()

        buInfo = wi.QPushButton("Info")
        buInfo.clicked.connect(self.info)
        gr.addWidget(buInfo, 0, 0)

        buWarning = wi.QPushButton("Warning")
        buWarning.clicked.connect(self.warning)
        gr.addWidget(buWarning, 0, 1)
```

```
buError = wi.QPushButton("Error")
buError.clicked.connect(self.error)
gr.addWidget(buError, 0, 2)

buYesNo = wi.QPushButton("Yes/No")
buYesNo.clicked.connect(self.yesno)
gr.addWidget(buYesNo, 1, 0)

buOkAbortRetryIgnore = wi.QPushButton(
    "Ok, Abort, Retry or Ignore")
buOkAbortRetryIgnore.clicked.connect(self.okabortretryignore)
gr.addWidget(buOkAbortRetryIgnore, 1, 1, 1, 2)

self.lbOutput = wi.QLabel("(empty)")
gr.addWidget(self.lbOutput, 2, 0)

buEnd = wi.QPushButton("End")
buEnd.clicked.connect(wi.QApplication.instance().quit)
gr.addWidget(buEnd, 2, 2)

self.setWindowTitle("Messages")
self.setLayout(gr)
self.show()
...
```

Listing 11.16 pyqt_message.py File, Part 1 of 3

The resulting GUI is shown in Figure 11.22. The wide button is positioned in the middle column. Thanks to the values of the last two parameters of the addWidget() method, this button extends over one row and two columns.

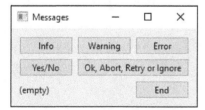

Figure 11.22 Buttons for Calling the Dialog Boxes

The second part of the program contains the calls of the three simple dialog boxes. The information contained in these boxes only needs to be confirmed.

...

```python
    def info(self):
        mb = wi.QMessageBox(self)
        mb.setWindowTitle("Box")
        mb.setIcon(wi.QMessageBox.Icon.Information)
        mb.setText("Info")
        mb.exec()

    def warning(self):
        mb = wi.QMessageBox(self)
        mb.setWindowTitle("Box")
        mb.setIcon(wi.QMessageBox.Icon.Warning)
        mb.setText("Warning")
        mb.exec()

    def error(self):
        mb = wi.QMessageBox(self)
        mb.setWindowTitle("Box")
        mb.setIcon(wi.QMessageBox.Icon.Critical)
        mb.setText("Error")
        mb.exec()
```

...

Listing 11.17 pyqt_message.py File, Part 2 of 3

The **Info**, **Warning**, and **Error** buttons lead to the info(), warning(), and error() methods, respectively. An object of the QMessageBox type is created in each case. A reference to the current object (i.e., the application window), is transferred as a parameter.

The setWindowTitle(), setIcon(), and setText() methods are used to display a title, an icon, and an information text in the dialog box. Icons are displayed using the Information, Warning, and Critical elements of the QMessageBox.Icon enumeration, as shown in Figure 11.23 through Figure 11.25.

The exec() method displays the dialog box. The entire application only continues to run once the dialog box has been closed.

Figure 11.23 Information Message, with Confirmation

Figure 11.24 Warning Message, with Confirmation

Figure 11.25 An Error Message, with Confirmation

11

The third and final part of our program contains the calls for the two dialog boxes in which a selection is made.

```
...
    def yesno(self):
        mb = wi.QMessageBox(self)
        mb.setWindowTitle("Box")
        mb.setIcon(wi.QMessageBox.Icon.Question)
        mb.setText("Yes or No")
        mb.setStandardButtons(wi.QMessageBox.StandardButton.Yes
            | wi.QMessageBox.StandardButton.No)
        mb.setDefaultButton(wi.QMessageBox.StandardButton.No)
        if mb.exec() == wi.QMessageBox.StandardButton.Yes:
            self.lbOutput.setText("Yes")
        else:
            self.lbOutput.setText("No")

    def okabortretryignore(self):
        mb = wi.QMessageBox(self)
        mb.setWindowTitle("Box")
        mb.setIcon(wi.QMessageBox.Icon.Question)
        mb.setText("Abort, Retry or Ignore")
        mb.setStandardButtons(wi.QMessageBox.StandardButton.Ok
            | wi.QMessageBox.StandardButton.Abort
            | wi.QMessageBox.StandardButton.Retry
            | wi.QMessageBox.StandardButton.Ignore)
        mb.setDefaultButton(wi.QMessageBox.StandardButton.Ignore)
        result = mb.exec()
```

```
            if result == wi.QMessageBox.StandardButton.Ok:
                self.lbOutput.setText("Ok")
            elif result == wi.QMessageBox.StandardButton.Abort:
                self.lbOutput.setText("Abort")
            elif result == wi.QMessageBox.StandardButton.Retry:
                self.lbOutput.setText("Retry")
            else:
                self.lbOutput.setText("Ignore")

application = wi.QApplication(sys.argv)
window = Window()
sys.exit(application.exec())
```

Listing 11.18 pyqt_message.py File, Part 3 of 3

The icon is displayed using the Question element of the QMessageBox.Icon enumeration.

Usually, only the **OK** button is displayed in a dialog box that is created using a QMessage-Box object.

The setStandardButtons() method is required to display other buttons. The buttons are created using the Yes, No, Ok, Abort, Retry, and Ignore elements of the QMessageBox.StandardButton enumeration. They are linked together using the bit operator |.

The setDefaultButton() method selects the button that has the focus once the dialog box has been called, as shown in Figure 11.26 and Figure 11.27.

The exec() method returns a reference to the selected button. In response, the corresponding text is displayed in the label in this program.

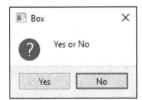

Figure 11.26 Answering "Yes" or "No"

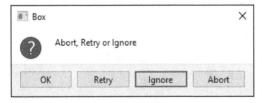

Figure 11.27 A Selection of Four Options

11.3 PyQt and SQLite

In this section, we want to create a GUI using PyQt. This application is designed to manage the SQLite database, *company.db*, which we introduced in Chapter 9, Section 9.2. This comprehensive application combines two important technologies and other elements of Python programming.

Using the program in the *pyqt_sqlite_admin.py* file, you can view all data records in a list widget and insert, change, and delete individual data records. You can also search for data records that meet specific criteria.

11.3.1 The User Interface of the Application

After launching the program, you can display all data records in a list widget, as shown in Figure 11.28. You can also edit the data of an individual data record in input fields.

Figure 11.28 Administering a Database

After pressing the various buttons, the following happens:

- **Display all button**
  All data records are displayed in the list widget.
- **List**
  After selecting a data record in the list widget, its individual data is displayed in the input fields.
- **Insert button**
  A data record is created from the data in the input fields and inserted into the database table.
- **Change button**
  The previously selected data record is updated with the data from the input fields.

- **Delete button**
  The selected data record displayed in the input fields is deleted after confirmation.

- **Search in name button**
  A search string can be entered in the **Name** input field. After clicking the button, all data records containing the search string in the Name field are displayed in the list widget.

11.3.2   Widgets on the Left

This extensive program is presented and explained in several sections. Let's look at the first part.

```python
import PyQt6.QtWidgets as wi
import os, sys, sqlite3

class Window(wi.QWidget):
    def __init__(self):
        super().__init__()
        gr = wi.QGridLayout()

        # Buttons in column 0
        buDisplayAll = wi.QPushButton("Display all")
        buDisplayAll.clicked.connect(self.displayAll)
        gr.addWidget(buDisplayAll, 0, 0)

        buInsert = wi.QPushButton("Insert")
        buInsert.clicked.connect(self.insert)
        gr.addWidget(buInsert, 1, 0)

        buChange = wi.QPushButton("Change")
        buChange.clicked.connect(self.change)
        gr.addWidget(buChange, 2, 0)

        buDelete = wi.QPushButton("Delete")
        buDelete.clicked.connect(self.delete)
        gr.addWidget(buDelete, 3, 0)
```

Listing 11.19 pyqt_sqlite_admin.py File, Part 1 of 13

The os, sys, and sqlite3 modules are required. The four buttons of type QPushButton in the column on the left call the displayAll(), insert(), change(), and delete() methods.

11.3.3   Widgets in the Middle

Now, let's move to the second part.

. . .

```
        # Label in column 1
        lbName = wi.QLabel("Name:")
        gr.addWidget(lbName, 1, 1)

        lbFirstName = wi.QLabel("First name:")
        gr.addWidget(lbFirstName, 2, 1)

        lbPersonnelnumber = wi.QLabel("Personnel number:")
        gr.addWidget(lbPersonnelnumber, 3, 1)

        lbSalary = wi.QLabel("Salary:")
        gr.addWidget(lbSalary, 4, 1)

        lbBirthday = wi.QLabel("Birthday:")
        gr.addWidget(lbBirthday, 5, 1)
. . .
```

Listing 11.20 pyqt_sqlite_admin.py File, Part 2 of 13

The five labels of type QLabel are used to label the input fields for the five fields of the
people table.

11.3.4   Widgets in the Right

The third part contains two buttons and five input fields.

. . .

```
        # Buttons and input fields in column 2
        buSearch = wi.QPushButton("Search in name")
        buSearch.clicked.connect(self.search)
        gr.addWidget(buSearch, 0, 2)

        buEnd = wi.QPushButton("End")
        buEnd.clicked.connect(wi.QApplication.instance().quit)
        gr.addWidget(buEnd, 7, 2)

        self.leName = wi.QLineEdit()
        gr.addWidget(self.leName, 1, 2)
```

```
        self.leFirstName = wi.QLineEdit()
        gr.addWidget(self.leFirstName, 2, 2)

        self.lePersonnelnumber = wi.QLineEdit()
        gr.addWidget(self.lePersonnelnumber, 3, 2)

        self.leSalary = wi.QLineEdit()
        gr.addWidget(self.leSalary, 4, 2)

        self.leBirthday = wi.QLineEdit()
        gr.addWidget(self.leBirthday, 5, 2)
...
```

Listing 11.21 pyqt_sqlite_admin.py File, Part 3 of 13

The **Search in name** button calls the search() method, while the **End** button ends the application. The five input fields are of type QLineEdit.

11.3.5   List Widget and Python List

Let's take a look at part 4.

```
...
        # List widget
        self.liDisplay = wi.QListWidget()
        self.liDisplay.setFixedHeight(80)
        self.liDisplay.itemClicked.connect(self.selection)
        gr.addWidget(self.liDisplay, 6, 0, 1, 3)

        # Python list parallel to the display in the list widget
        self.pnumber = []
...
```

Listing 11.22 pyqt_sqlite_admin.py File, Part 4 of 13

The list widget of the QListWidget type is positioned in the column on the left. Due to the values of the last two parameters of the addWidget() method, the list widget extends over one row and three columns. The pnumber Python list is initially empty. Over the course of the program, it is ensured that the Python list always contains a list of the personnel numbers of those data records that are currently displayed in the list widget, in the same order. In this way, the associated data records in the database can be accessed.

11.3.6   Database File and Window

Access to the SQLite database file follows in part 5.

...

```
        # Check database file, create if necessary
        if os.path.exists("company.db"):
            try:
                connection = sqlite3.connect("company.db")
                cursor = connection.cursor()
                sql = "SELECT * FROM people"
                cursor.execute(sql)
            except sqlite3.Error as e:
                self.error(e)
        else:
            try:
                connection = sqlite3.connect("company.db")
                cursor = connection.cursor()
                sql = "CREATE TABLE people(name TEXT, firstname TEXT," \
                    " personnelnumber INTEGER PRIMARY KEY, salary REAL," \
                    " birthday TEXT)"
                cursor.execute(sql)
            except sqlite3.Error as e:
                self.error(e)
        connection.close()

        # Design and display window
        self.setWindowTitle("Database, Administration")
        self.setLayout(gr)
        self.show()
```

...

Listing 11.23 pyqt_sqlite_admin.py File, Part 5 of 13

The program checks whether the database file *company.db* exists in the same directory as the application:

- If the file exists, the people table will be accessed to determine whether a database error would occur during access.
- If it does not exist, the database file, the database, and the people table will be created.

If a database error occurs during one of these actions, the corresponding error message is output using the error() method described in the next section. The sqlite3.Error class is the base class for the exception classes in connection with accessing SQLite databases.

A title and the filled-in grid layout are assigned to the application window, which is then displayed. At this point, the constructor method of the Window class ends.

11.3.7   Three Auxiliary Methods

Part 6 describes the three auxiliary methods error(), message(), and input_empty(), each of which is required at several points in the program.

...

```
    # Auxiliary method: Display of an SQLite error
    def error(self, e):
        mb = wi.QMessageBox(self)
        mb.setWindowTitle("Error")
        mb.setText(str(e))
        mb.exec()

    # Auxiliary method: Display of an info message
    def message(self, info):
        mb = wi.QMessageBox(self)
        mb.setWindowTitle("Info")
        mb.setText(info)
        mb.exec()

    # Auxiliary method: Empty input fields
    def inputFieldsEmpty(self):
        self.leName.setText("")
        self.leFirstName.setText("")
        self.lePersonnelnumber.setText("")
        self.leSalary.setText("")
        self.leBirthday.setText("")
```

...

Listing 11.24 pyqt_sqlite_admin.py File, Part 6 of 13

When the error() auxiliary method is called, an object of an Exception class is transferred. The str() function returns text information for the transferred object of the Exception class. An error message is displayed using an object of the QMessageBox type, as shown in Figure 11.29.

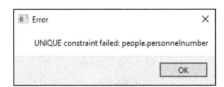

Figure 11.29 Error Message from the Database

When the `message()` auxiliary method is called, information is transferred that is also displayed using an object of the `QMessageBox` type, as shown in Figure 11.30.

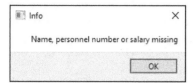

Figure 11.30 Information from the Application

The process of the `input_empty_fields()` auxiliary method is self-explanatory. Once all data records have been displayed, the input fields should not contain any "old" data that was previously used.

11.3.8 Display in the List Widget

After clicking the **Display all** button, the `displayAll()` method is called. Let's tackle this part next.

```
...
    # "Display all" button
    def displayAll(self):
        self.inputFieldsEmpty()
        self.liDisplay.clear()
        self.pnumber.clear()

        try:
            connection = sqlite3.connect("company.db")
            cursor = connection.cursor()
            sql = "SELECT * FROM people"
            cursor.execute(sql)
            for dr in cursor:
                self.liDisplay.addItem(f"{dr[0]}
                    # {dr[1]}" \f"
                    # {dr[2]} # {dr[3]} # {dr[4]}")
                self.pnumber.append(dr[2])
            if len(self.pnumber) == 0:
                self.message("No data record")
            self.liDisplay.setCurrentRow(0)
        except sqlite3.Error as e:
            self.error(e)
        connection.close()
...
```

Listing 11.25 pyqt_sqlite_admin.py File, Part 7 of 13

In the displayAll() method, the input fields, the list widget, and the pnumber Python list are cleared first.

All data records in the people table are retrieved. In a loop, both the data records are added to the list widget and the corresponding personnel numbers are added to the Python list. The data in a data record is separated by a character string containing the # character.

If the table does not contain a data record, a corresponding message will be displayed. If a database error occurs, the corresponding error message is displayed using the error() method.

11.3.9   Selection in a List Widget

After selecting an entry in the list widget, the select() method is called in the next part.

```
...
    # Selection in list widget
    def selection(self):
        i = self.liDisplay.currentRow()
        if i == -1:
            self.message("No data record selected")
            return
        try:
            connection = sqlite3.connect("company.db")
            cursor = connection.cursor()
            sql = "SELECT * FROM people WHERE personnelnumber = " \
                + str(self.pnumber[i])
            cursor.execute(sql)
            for dr in cursor:
                self.leName.setText(dr[0])
                self.leFirstName.setText(dr[1])
                self.lePersonnelnumber.setText(str(dr[2]))
                self.leSalary.setText(str(dr[3]))
                self.leBirthday.setText(dr[4])
        except sqlite3.Error as e:
            self.error(e)
        connection.close()
...
```

Listing 11.26 pyqt_sqlite_admin.py File, Part 8 of 13

The currentRow() method returns the number of the selected entry in the list widget. The first entry has the number 0. If no entry is selected, the value -1 is used. A data record can only be displayed if it was previously selected in the list widget.

Thanks to the unique personnel numbers, the result of the query comprises exactly one data record. The selection is made using the Python list that was created in parallel to the content of the list widget. The data of the data record is displayed in the input fields.

11.3.10   Inserting a New Data Record

The **Insert** button leads to the insert() method, which we'll set up next.

```
...
    # "Insert" button
    def insert(self):
        if self.leName.text() == "" or self.lePersonnelnumber.text() == "" \
                or self.leSalary.text() == "":
            self.message("Name, personnel number or salary missing")
            return
        try:
            connection = sqlite3.connect("company.db")
            cursor = connection.cursor()
            sql = "INSERT INTO people VALUES(?, ?, ?, ?, ?)"
            cursor.execute(sql, (self.leName.text(), self.leFirstName.text(),
                int(self.lePersonnelnumber.text()), float(self.leSalary.text()),
                self.leBirthday.text())))
            connection.commit()
        except sqlite3.Error as e:
            self.error(e)
        except Exception as e:
            self.error(e)
        connection.close()
        self.displayAll()
...
```

Listing 11.27 pyqt_sqlite_admin.py File, Part 9 of 13

A new data record can only be inserted if at least the name, the personnel number, and the salary are available.

The data from the input fields is embedded in the SQL statement using parameter substitution to prevent malicious SQL code (*SQL injection*), see Chapter 9, Section 9.2.6.

The string with the entered personnel number is converted into an integer, while the string with the entered salary is converted into a number with decimal places.

If an existing personnel number is used or another database error occurs, the corresponding error message is displayed (shown earlier in Figure 11.29). If an error occurs during the conversion of the strings into numbers, a corresponding error message also appears thanks to the second exception branch, as shown in Figure 11.31.

At the end, all current data records are output again.

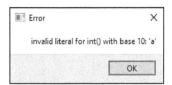

Figure 11.31 Error During Conversion

11.3.11   Changing a Data Record

The **Change** button leads to the change() method, as illustrated in the next part.

```
...
    # "Change" button
    def change(self):
        i = self.liDisplay.currentRow()
        if i == -1:
            self.message("No data record selected")
            return
        try:
            connection = sqlite3.connect("company.db")
            cursor = connection.cursor()
            sql = "UPDATE people SET name = ?, firstname = ?," \
                " personnelnumber = ?, salary = ?, birthday = ?" \
                " WHERE personnelnumber = " + str(self.pnumber[i])
            cursor.execute(sql, (self.leName.text(), self.leFirstName.text(),
                int(self.lePersonnelnumber.text()), float(self.leSalary.text()),
                self.leBirthday.text())))
            connection.commit()
        except sqlite3.Error as e:
            self.error(e)
        except Exception as e:
            self.error(e)
        connection.close()
        self.displayAll()
...
```

Listing 11.28 pyqt_sqlite_admin.py File, Part 10 of 13

A data record can only be changed if it was previously selected in the list widget.

In this case too, the data from the input fields is embedded in the SQL statement using parameter substitution. The data record to be updated is selected using the Python list that is maintained in parallel. The original personnel number is stored in it, but changing the personnel number is possible.

The conversions, the error messages, and the final output run in the same way as when you insert a new data record.

11.3.12 Deleting a Data Record

The **Delete** button leads to the delete() method, which we'll cover next.

```
...
    # "Delete" button
    def delete(self):
        i = self.liDisplay.currentRow()
        if i == -1:
            self.message("No data record selected")
            return

        # Query
        mb = wi.QMessageBox(self)
        mb.setWindowTitle("Query")
        mb.setText("Sure to delete?")
        mb.setStandardButtons(wi.QMessageBox.StandardButton.Yes
            | wi.QMessageBox.StandardButton.No)
        mb.setDefaultButton(wi.QMessageBox.StandardButton.No)
        if mb.exec() == wi.QMessageBox.StandardButton.No:
            return

        # Delete
        try:
            connection = sqlite3.connect("company.db")
            cursor = connection.cursor()
            sql = "DELETE FROM people WHERE personnelnumber = " \
                + str(self.pnumber[i])
            cursor.execute(sql)
            connection.commit()
        except sqlite3.Error as e:
            self.error(e)
        connection.close()
        self.displayAll()
...
```

Listing 11.29 pyqt_sqlite_admin.py File, Part 11 of 13

A data record can only be deleted if it was previously selected in the list widget. To be on the safe side, a query is made using a dialog box before deletion, as shown in Figure 11.32.

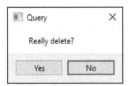

Figure 11.32 Query

The deletion is only carried out if the **Yes** button has been selected. The data record to be deleted is selected using the parallel Python list.

The error messages and the final output are the same as when you insert a new data record.

11.3.13   Searching for a Name

In the next part, the **Search in name** button leads to the search() method.

```
...
    # "Search in name" button
    def search(self):
        entry = self.leName.text()
        self.inputFieldsEmpty()
        self.liDisplay.clear()
        self.pnumber.clear()

        try:
            connection = sqlite3.connect("company.db")
            cursor = connection.cursor()
            sql = "SELECT * FROM people WHERE name LIKE ?"
            entry = '%' + entry + '%'
            cursor.execute(sql, (entry,))
            for dr in cursor:
                self.liDisplay.addItem(f"{dr[0]} # {dr[1]}" \
                    f" # {dr[2]} # {dr[3]} # {dr[4]}")
                self.pnumber.append(dr[2])
            self.liDisplay.setCurrentRow(0)
            if len(self.pnumber) == 0:
                self.message("No matching data record")
        except sqlite3.Error as e:
```

```
        self.error(e)
    connection.close()
...
```

Listing 11.30 pyqt_sqlite_admin.py File, Part 12 of 13

The content of the input field for the name is saved. This content can be a full name or just part of a name. The input fields, the list widget, and the pnumber Python list are then cleared.

The list widget displays all data records in the people table that contain the previously saved content of the input field. The content is embedded in percentage signs and inserted into the SQL statement using parameter substitution. The pnumber Python list is filled in parallel to the list widget. If no data record is found that matches the content of the input field, a corresponding message will display.

11.3.14   Main Program

Finally, we've come to the main program, which is rather short.

```
...
# Start and exit the application
application = wi.QApplication(sys.argv)
window = Window()
sys.exit(application.exec())
```

Listing 11.31 pyqt_sqlite_admin.py File, Part 13 of 13

In this case, the PyQt application is started and terminated again after clicking the **End** button.

11.4   Exercise

u_pyqt_language
Use the *PyQt* interface in the *u_pyqt_language.py* file to create a simple language learning program based on an SQLite database. The logic, structure, and user interface correspond to similar elements from the u_gui_language exercise in Chapter 10, Section 10.8, as shown in Figure 11.33.

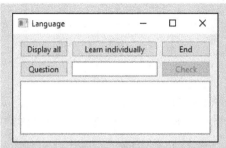

Figure 11.33 GUI after the Call

If you're not yet familiar with the development of the program, you'll find some tips in Appendix A, Section A.5.

Appendix A
Miscellaneous Topics

This appendix contains instructions for installing additional modules, creating executable EXE files, and installing the XAMPP package for various operating systems. It follows a compilation of important UNIX commands that you can use on Ubuntu Linux and macOS.

A.1 Installing Additional Modules

For a number of programs in this book, additional modules for Python must be installed first.

A.1.1 Installation on Windows

On Windows, the *pip* package management program is required and should already have been installed together with Python. To use it, open a command prompt (see Chapter 2, Section 2.3.2) and go to the *Scripts* subdirectory using the corresponding cd commands.

Then, you can install an additional module. For example, you can install matplotlib with the following command:

```
pip install matplotlib
```

A.1.2 Installation on Ubuntu

On Ubuntu Linux, an additional module is installed using the following call, again using matplotlib, as an example:

```
sudo apt install python3-matplotlib
```

The python3 name is followed by a hyphen and then the name of the module. The *mypy* type testing program is an exception; it is installed with the following command:

```
sudo apt install mypy
```

A.1.3 Installation on macOS

The *pip* package management program is required on macOS, so you should first download and install it. This step requires the following two calls in the terminal:

```
curl https://bootstrap.pypa.io/get-pip.py -o get-pip.py
python3 get-pip.py
```

To use it, open a terminal (see Chapter 2, Section 2.3.3). Then, you can install an additional module, again using matplotlib, as an example:

```
pip install matplotlib
```

A.2 Creating Executable Files

Using the pyinstaller program, you can create an executable file from any of your Python programs on Windows or macOS. This concept is also true for graphical user interface (GUI) programs.

The executable file can be started on another computer with the same operating system. It is not necessary to install Python on the other computer beforehand. To create an executable version of the *hello.py* program, follow these steps:

1. First, install the pyinstaller program using the *pip* package management program (Section A.1) via the following call: pip install pyinstaller.
2. On Windows: Copy the *hello.py* file to the *Scripts* subdirectory of your Python installation (i.e., to *C:\Python\Scripts*); open a command prompt (see Chapter 2, Section 2.3.2); and go to the *Scripts* subdirectory using the corresponding cd commands.
3. On macOS: Copy the *hello.py* file to the directory you want to use; open a terminal (see Chapter 2, Section 2.3.3); and go to the selected directory.
4. Call the pyinstaller program with your Python program, in this case, with pyinstaller hello.py.
5. In the *dist* subdirectory you'll then find the *hello* subdirectory including the executable *hello.exe* or the *hello* file.
6. Copy this directory completely to the target computer and call the executable *hello.exe* or *hello* file from the command line. On macOS, this call is performed using ./hello.

If you do not call the executable file from the command line, the program will also start but then will end quickly. The output window would only stop if an input were requested at the end of the program, for example using input().

A.3   Installing XAMPP

I will now describe the installation of the freely available, preconfigured XAMPP package, which you can use to access a MySQL database server (or its spin-off MariaDB) on Windows, Ubuntu Linux, or macOS. This package also includes an Apache web server, the PHP language, and the phpMyAdmin database GUI. The package is currently (as of spring 2024) available in version 8.2.12. Note that you must adjust the name of the executable file, as described in the following sections.

Download the XAMPP package for your operating system from *https://www.apache-friends.org* and navigate to your download directory.

A.3.1   Installing XAMPP on Windows

The XAMPP package for Windows can be found in the executable file *xampp-windows-x64-8.2.12-0-VS16-installer*. Start the installation by calling that file. In some cases, two warnings appear at the beginning, including one about a running antivirus program. You can continue by confirming the suggested installation options. I recommend selecting *C:\xampp* as the target directory. After the installation, start the **XAMPP Control Panel** application. Start and stop the MariaDB database server using the button to the right of the **MySQL** term.

A.3.2   Installing XAMPP on Ubuntu Linux

The XAMPP package for Ubuntu Linux can be found in the *xampp-linux-x64-8.2.12-0-installer.run* file. Open a terminal for input. If necessary, change the access rights to the file by using the following command:

```
chmod 744 xampp-linux-x64-8.2.12-0-installer.run
```

Start the installation with the following command:

```
sudo ./xampp-linux-x64-8.2.12-0-installer.run
```

Then, confirm the suggested installation options. XAMPP will be installed in the */opt/lampp* directory. At the end of the installation process, you can leave the **Launch Xampp** checkbox selected. This option opens a dialog box for managing the servers. Under the **Manage Servers** tab, you have the option of selecting the MariaDB database server (via **MySQL**) and starting and stopping it using the button on the right. You can also call the dialog box for managing the servers directly with the following command:

```
sudo /opt/lampp/manager-linux-x64.run
```

A.3.3   Installing XAMPP on macOS

You can download the *XAMPP for OS X* package from *https://www.apachefriends.org*. The *xampp-osx-8.2.12-0-installer.dmg* file is then available. Double-click on this file to create a new drive. Then, you can call the installation file located on the new drive.

As the installation program does not come from Apple itself, this executable is initially blocked by macOS. Open the **Security** menu item in the system settings. You'll see the reference to XAMPP and the **Open anyway** button. After clicking the button, you have the option of starting the installation program.

XAMPP is installed by default in the */Applications/XAMPP* directory, which corresponds to the *Applications/XAMPP* directory in the **Finder**.

At the end of the installation, you can leave the **Launch Xampp** checkbox selected. This setting opens a dialog box for managing various servers. Under the **Manage Servers** tab, you have the option of selecting the MariaDB database server (via **MySQL**) and starting and stopping it using the button on the right. You can also call the dialog box for managing the various servers via *Programs/XAMPP/manager-osx*.

A.4   UNIX Commands

To manage directories and files on the UNIX operating system or one of its descendants, such as Ubuntu Linux or macOS, you can use command line commands. These commands can be entered in a terminal on these operating systems.

In this section, you'll learn about a few useful commands: ls, mkdir, rmdir, cd, cp, mv, and rm. Be sure you pay attention to the difference between uppercase and lowercase letters. For example, you might have two different files named *hello.txt* and *Hello.txt*.

A.4.1   Contents of a Directory

Like Windows systems, UNIX systems have hierarchies of directories. For this reason, you have a main directory, under which you may have subdirectories, which in turn may have other subdirectories.

By using .. (two dots) you can always address the parent directory, while . (a single dot) refers to the current directory. These designations are also used for some commands.

Use the ls -l command to display a detailed list of the files and subdirectories in the current directory. This command can be a useful check after every change. Here is a sample output:

```
-rw-r--r--  1  theis theis    12  Dec 3  08:52  hello.txt
-rw-r--r--  1  theis theis    12  Dec 3  08:51  greetings.txt
drwxr-xr-x  2  theis theis  4096  Dec 3  08:57  house
```

In this example, we can see two files with the extension *.txt* and a subdirectory. The following information is most important:

- In contrast to Windows, the access rights are clearly divided. Not everyone is allowed to do everything. If you see a d (for *directory*) in the first column, this entry is a subdirectory.

- This information is followed by the rights relating to the entry. In this case, r (for *read*) stands for read permission, w (for *write*) for write permission, and x (for *execute*) for the right to execute a program.

- These rights are listed three times in succession: first for the current user, then for their workgroup, and finally for everyone.

- The sizes of the files are displayed. In this case, we have 12 bytes each in the two text files. You can also see the date and time of the last change.

A.4.2   Creating, Changing, and Deleting a Directory

You can use the mkdir command (*make directory*) to create a new directory below the current directory. For example:

- mkdir myTexts creates the myTexts subdirectory relative to the current directory.

The cd command (*change directory*) enables you to change the directory. Let's consider some examples:

- cd myTexts changes to the *myTexts* subdirectory relative to the current directory.
- cd .. changes to the parent directory.
- cd (without further details) changes to your home directory, regardless of the current directory.
- cd /usr/bin changes to the absolute *usr/bin* directory, independent of the current directory.

You can use the rmdir command (*remove directory*) to delete an empty subdirectory that is located below the current directory. To delete files from a directory, refer to the next section. For example:

- rmdir myTexts deletes the myTexts subdirectory if that subdirectory is empty.

A.4.3   Copying, Moving, and Deleting Files

You can create text files using a text editor such as *gedit*, *vi*, or *nano*. Python programs are best created within the Integrated Development and Learning Environment (IDLE).

To copy files, you can use the cp command (*copy*). In this context, two entries are always required, one for the source and one for the destination of the copying process. Let's look at some examples:

- cp hello.txt greetings.txt copies the *hello.txt* file to the *greetings.txt* file within the current directory.
- cp hello.txt .. copies the *hello.txt* file to the parent directory.
- cp hello.txt ../evenMoreTexts copies the *hello.txt* file to the *evenMoreTexts* directory, which is located under the same parent directory as the current directory.
- cp ../hello.txt . copies the *hello.txt* file from the parent directory to the current directory. (Note the single dot after the space at the end of the command.)
- cp ../*.txt . copies all files with the *.txt* extension from the parent directory to the current directory.
- cp ../evenMoreTexts/hello.txt . copies the *hello.txt* file from the *evenMoreTexts* directory (see earlier) to the current directory.

The mv command (*move*) allows you to rename or move files. Its use is similar to that of cp. Let's look at some examples:

- mv hello.txt greetings.txt renames the *hello.txt* file to *greetings.txt* within the current directory.
- mv hello.txt .. moves the *hello.txt* file to the parent directory.
- mv ../hello.txt . moves the *hello.txt* file from the parent directory to the current directory.
- mv ../*.txt . moves all files with the *.txt* extension from the parent directory to the current directory.

To delete files, you can use the rm command (*remove*). Let's look at some examples:

- rm hello.txt deletes the *hello.txt* file within the current directory.
- rm *.txt deletes all files with the *.txt* extension within the current directory.

A.5   Tips for Development

The following are tips for developing specific exercises.

A.5.1   Exercise u_sqlite_language

This exercise is described in Chapter 9, Section 9.2.9.

If the SQLite database *language.db* exists at the start of the program, its contents are loaded into a two-dimensional list in a function named load(). A reference to the list is provided as the return value of the function.

If the SQLite database does not yet exist at the start of the program, a two-dimensional list is filled in with the data in the create() function. The *language.db* database is then created, and then the language table is created containing the id, German, and English fields. The contents of the list are then saved in the database so that the database is available the next time the program is launched. A reference to the list is provided as the return value of the function.

The menu is integrated in an outer loop that can only be exited by entering "0". After successfully passing the test, the program must be terminated. After entering a "1", "2", or "3", functions named all(), learn(), and test() will be called, respectively. During the call, a reference to the two-dimensional list is passed as a parameter.

The input of the desired menu item is embedded in an inner loop that can only be exited by entering a value from "0" to "3". A different value leads to the prompt being repeated.

A.5.2   Exercise u_gui_language

This exercise is described in Chapter 10, Section 10.8.

The logic of the program is the same as in exercise u_sqlite_language. The names and tasks of most functions also match.

The create() and load() functions can be adopted from the previous exercise as they are processes that are independent of the interface used.

The all() and learn() functions must be adapted to the GUI. The task of the test() function is split between the question() and check() functions. There is also an end() function.

When the last five functions are called, a reference to the two-dimensional list is passed as a parameter in each case. Because these functions are the callback functions for the buttons, a lambda function is required in each case.

In the question() function, the number of the currently selected German word is saved in a global variable. This variable is accessed in the check() function so that the corresponding English word can be used for comparison.

A.5.3   Exercise u_pyqt_language

This exercise is described in Chapter 11, Section 11.4.

The create() and load() functions can be used again. In this case, these functions are called in the constructor of the Window class, and each return a reference to the two-dimensional list. This result is saved as a property of the Window class.

In this way, the list is available in all other methods of the window class. No reference to the list needs to be passed when the methods are called.

In the question() method, the number of the currently selected German word is saved in a property of the Window class. This property is accessed in the check() function so that the corresponding English word can be used for comparison.

Appendix B
The Author

 **Thomas Theis** has more than 40 years of experience as a software developer and as an IT lecturer. He holds a graduate degree in computer engineering. He has taught at numerous institutions, including the Aachen University of Applied Sciences. He conducts training courses for JavaScript, C/C++, Visual Basic, and web programming, and is the author of several successful technical books.

Index

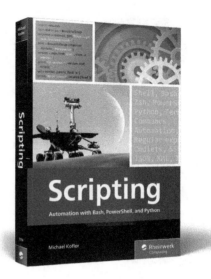

- Learn to work with scripting languages such as Bash, PowerShell, and Python

- Get to know your scripting toolbox: cmdlets, regular expressions, filters, pipes, and REST APIs

- Automate key tasks, including backups, database updates, image processing, and web scraping

Michael Kofler

Scripting

Automation with Bash, PowerShell, and Python

Developers and admins, it's time to simplify your workday. With this practical guide, use scripting to solve tedious IT problems with less effort and fewer lines of code! Learn about popular scripting languages: Bash, PowerShell, and Python. Master important techniques such as working with Linux, cmdlets, regular expressions, JSON, SSH, Git, and more. Use scripts to automate different scenarios, from backups and image processing to virtual machine management. Discover what's possible with only 10 lines of code!

470 pages, pub. 02/2024
E-Book: $44.99 | **Print:** $49.95 | **Bundle:** $59.99

www.rheinwerk-computing.com/5851